Automation, Innovation and Economic Crisis

The fourth industrial revolution is developing globally, with no geographical centre. It is also taking place at enormous speed. This development will shape the workplaces of the future, which will be entirely different from the workplaces created by the first, second and third industrial revolutions. Industry created the industrial worker. The knowledge society will create a new type of "industrial worker", the knowledge worker. While the third industrial revolution was concerned with the digitalization of work, in the fourth industrial revolution, robots will bring about the informatization of work. Many of these robots will be systematically connected, such that they can obtain updated information and learn from their own and others' mistakes. The way we work, where we work, what we work on, and our relationships with our colleagues and employers are all in a state of change. The workplace of the future will not necessarily be a fixed geographical location, but may be geographically distributed and functionally divided.

In his book, Jon-Arild Johannessen argues that a "perfect" social storm occurs when inequality grows at a catastrophic rate, unemployment increases, job security is threatened for a growing number and robotization takes over even the most underpaid jobs. Thus, the ingredients for a perfect social storm will be brought forward by cascades of innovations that will most likely lead to economic and social crises and he argues that it is reasonable to assume that it will only take a small spark for this social storm to develop into a social revolution.

Jon-Arild Johannessen holds a Master of Science from Oslo University in History. He holds a Ph.D. from Stockholm University in Systemic Thinking. He is currently professor (full) in Leadership, at Kristiania University College, Oslo, and Nord University, Norway. He has been professor (full) in Innovation at Syd-Danske University, Denmark. He has been professor (full) in Management at The Arctic University, Norway. At Bodø Graduate School of Business, Norway, he had a professorship (full) in Information Management. At the Norwegian School of Management he has been professor (full) in Knowledge Management. He has written more than 200 international papers on innovation, knowledge management and leadership. He has been author/co-author of 30 books.

Routledge Studies in the Economics of Innovation

The Routledge Studies in the Economics of Innovation series is our home for comprehensive yet accessible texts on the current thinking in the field. These cutting-edge, upper-level scholarly studies and edited collections bring together robust theories from a wide range of individual disciplines and provide in-depth studies of existing and emerging approaches to innovation, and the implications of such for the global economy.

Automation, Innovation and Economic Crisis
Surviving the Fourth Industrial Revolution
Jon-Arild Johannessen

The Economic Philosophy of the Internet of Things
James Juniper

For more information about this series, please visit: www.routledge.com/Routledge-Studies-in-the-Economics-of-Innovation/book-series/ECONINN

Automation, Innovation and Economic Crisis

Surviving the Fourth Industrial Revolution

Jon-Arild Johannessen

Routledge
Taylor & Francis Group
LONDON AND NEW YORK

First published 2018
by Routledge
2 Park Square, Milton Park, Abingdon, Oxon OX14 4RN

and by Routledge
711 Third Avenue, New York, NY 10017

Routledge is an imprint of the Taylor & Francis Group, an informa business

British Library Cataloguing in Publication Data
A catalogue record for this book is available from the British Library

Library of Congress Cataloging in Publication Data
Names: Johannessen, Jon-Arild, author.
Title: Automation, innovation and economic crisis : surviving the fourth industrial revolution / Jon-Arild Johannessen.
Description: Abingdon, Oxon ; New York, NY : Routledge, 2018. | Series: Economics of innovation | Includes bibliographical references and index.
Identifiers: LCCN 2018001648 | ISBN 9781138488601 (hardback) | ISBN 9781351039864 (ebook)
Subjects: LCSH: Robots--History--21st century. | Globalization--History--21st century. | Technological innovations--History--21st century.
Classification: LCC TJ211 .J544 2018 | DDC 330.9001/12--dc23
LC record available at https://lccn.loc.gov/2018001648

ISBN: 978-1-138-48860-1 (hbk)
ISBN: 978-1-351-03986-4 (ebk)

Typeset in Times New Roman
by Taylor & Francis Books

Contents

Figures

Acknowledgements

I want to thank the staff at Routledge, who helped me to make this book much better than I could have done by myself. Especially I want to thank the editor, Kristina Abbotts.

Preface

The book is divided into four chapters and a chapter on concepts used in the book. In Chapter 1 we discuss the question: What constitutes the fourth industrial revolution? The objective with the question is to understand the fourth industrial revolution. The findings in Chapter 1 are an analytical model that shows aspects of the factors which constitute the fourth industrial revolution, together with some of the key consequences of the fourth industrial revolution.

The question discussed in Chapter 2 is: How does globalization constitute an aspect of the fourth industrial revolution? The objective is to understand and explain how globalization affects, and is affected by, the fourth industrial revolution. The findings in Chapter 2 are that the global power system is fostering a totalitarian form of capitalism and controlling the global knowledge economy via financial capital in a new nation, "Mamounia", which consists of the One Per Cent of the population in the world.

The question discussed in Chapter 3 is: How does robotization constitute an aspect of the fourth industrial revolution? The objective is to conceptualize some of the consequences of the fourth industrial revolution. The findings in Chapter 3 are linked to two conceptual models explaining the question studied and the objective of the chapter.

Due to extreme economic inequality, the fourth industrial revolution may develop in such a way that it could be detrimental to organizations and nations. The question discussed in Chapter 4 is: Which overall policy can be developed so that the fourth industrial revolution will lead to greater value creation for the benefit of an increasing number of people?

The objective is to design a policy architecture that can reduce the development of strong tensions and conflicts in the fourth industrial revolution. The findings in Chapter 4 are 17 policy interventions that are systemically linked. This comprises six policy interventions for the economic subsystem; three policy interventions for the political subsystem; three policy interventions for the cultural subsystem and five policy interventions for the collaborative system.

1 Trends towards feudal capitalism

Introduction

The most important invention of the first industrial revolution was the steam engine. The second industrial revolution spawned the combustion engine, and the third industrial revolution the computer. The fourth industrial revolution will see the rise of robots. These robots will be integrated into all types of products, both known and unknown (Valnazarova & Ydesen, 2016).

The fourth industrial revolution is being driven forward by robotization and globalization (Schwab, 2016). Robots with artificial intelligence are of a qualitatively different order than the wave of automated machines we saw in the third industrial revolution (Brynjolfsson & Saunders, 2013). These robots can be programmed to sense their surroundings, understand what is happening, adapt their behaviour in accordance with this understanding and learn from their actions, without any need for reprogramming (Brynjolfsson & McAfee, 2011; 2014).

Obviously, we do not know how the fourth industrial revolution will progress, but we can apply historical insight and innovation theory, as well as our understanding of economics, in order to predict some likely developments.

The fourth industrial revolution is developing globally, with no geographical centre (Dickinson, 2016). It is also taking place at enormous speed. The spread of computers from the 1950s onwards was limited by many factors, including the world situation, East–West tensions and the absence of a well-developed internet (Rogers, 1962). No such limiting factors apply to the spread of robot technology. This spread is exponential, is taking place worldwide and is systemically connected (Reinert & Rogoff, 2009).

The fourth industrial revolution can be summed up in six concepts. We have described three of them above: the rate of spread, which is driven by cascading innovations; the global area of impact, which is driven by globalization; the systemic links, which are driven by robotization and informatization. The three other concepts are linked to the three concepts listed above. These concepts are threshold value, feedback and time-lag. Threshold value is relevant because some sectors will join the fourth industrial revolution only once a certain threshold value has been exceeded, not unlike the situation where a dam bursts once a specific threshold is exceeded. Feedback may be understood as meaning that some people understand what is about to happen before others. The former group react and adapt to the new development before others do so. Accordingly, the former group gain an advantage over the others, who will come on board later. Time-lag means that the more one is part of the fourth industrial revolution, the less of a time-lag one will be willing to accept. For example, this time-lag may be the time that elapses between sending a message and receiving a response. One of the consequences of this is that a relatively long time-lag will function psychologically as information stress.

These six concepts – rate of spread, global impact area, systemic links, threshold value, feedback and time-lag – function both as descriptions of what is happening in the fourth industrial revolution, and are also used here as explanations of the consequences of the fourth industrial revolution.

The knowledge society is mushrooming out of the global economy, and artificial intelligence and robotization are the key drivers (Frankish, 2014). This development will shape the workplaces of the future, which will be completely different from the workplaces created by the industrial revolution (Barrat, 2015). Industry created the industrial worker. The knowledge society will create a new type of "industrial worker", the knowledge worker (Case, 2016). The knowledge worker is not a homogeneous category, but consists of different types of worker, distinguished by level of competence, type of function performed and manner of thinking (Brynjolfsson & McAfee, 2014).

The five elements that emerge here – the knowledge society; globalization; robotization; the future workplace; and the knowledge worker – are all based on a revolutionary innovation that will change the basic assumptions underlying all social systems (Brynjolfsson & McAfee, 2011). Innovation is the overarching driver of the new knowledge society that we are seeing foreshadowed today, according to Brynjolfsson and Saunders (2013).

This chapter explores the following research problem: What constitutes the fourth industrial revolution?

The questions that we investigate in order to answer this research problem are the following:

- How does the knowledge society constitute an aspect of the fourth industrial revolution?
- How does globalization constitute an aspect of the fourth industrial revolution?
- How does robotization constitute an aspect of the fourth industrial revolution?
- How does the workplace of the future constitute an aspect of the fourth industrial revolution?

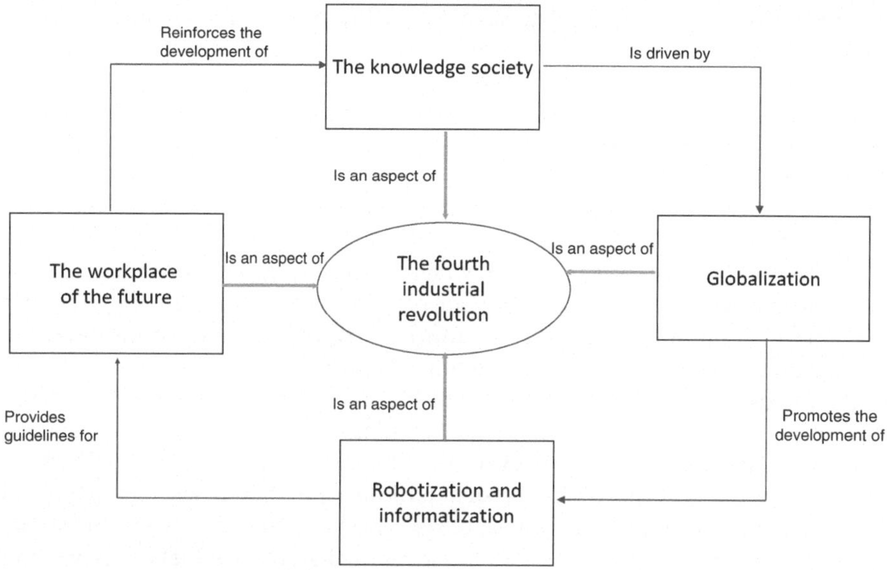

Figure 1.1 The fourth industrial revolution

This introduction is visualized in Figure 1.1, which also shows how we have structured this chapter.

Analysis and discussion

In the following, the four elements in Figure 1.1 will be analysed and discussed.

The knowledge society

Several simultaneous trends characterize developments towards the knowledge society. Innovation is without doubt one such tendency that will increase (Gans, 2016). However, it is debated whether the market alone will manage to accommodate the major challenges we face (Ford, 2016: 283). These challenges include migration from areas of famine, drought, poverty and an unworthy life, and people fleeing war, crises and conflicts (Johnson, 2014). Another challenge also related to migration is the growing climate crisis (Smil, 2012). A third challenge is the rising unemployment among people who lack the necessary education and expertise demanded of the fourth industrial revolution (Ford, 2016). A fourth challenge is the issue of economic disparity, the ever-widening gap between rich and poor, both globally and nationally, as well as in urban areas (Piketty, 2016).

Economic, psychological and social security will come under strong pressure (Bauman, 2013; Sennett, 1999, 2003, 2013). The use of terms such as "the working poor" (Shipler, 2005) and "the precariat" (Standing, 2014b; Savage, 2015) signal clear indications of this trend. The middle class, the backbone of capitalism in the 20th century, will be marginalized and decimated with the emergence of the fourth industrial revolution in the 21st century (Wacquant, 2007, 2009a, 2009b; Savage, 2015; Standing, 2014b). Even those with a long university education will experience job insecurity – many will have educated themselves into unemployment (Gupta et al., 2016; Coates, 2016).

A "perfect" social storm occurs when inequality grows at a catastrophic rate, unemployment increases, job security is threatened for a growing number and robotization takes over even the most underpaid jobs in, for instance, the service industries (Ford, 2016; Wacquant, 2009b).

Paradoxically, productivity increases for those who have jobs, due to the new robotic technology, and those working in high-tech manufacturing will enjoy higher wages. However, the conflict between those with well-paying jobs in high-productivity occupations, and those with poorly paid jobs and the unemployed, will increase sharply (Savage, 2015; Standing, 2014b). In addition, social tensions will increase due to a gradual disconnection between productivity and pay (Pilger, 2016). Thus, although productivity and profits increase, wages relatively decrease (Piketty, 2014; Ford, 2016). If this gradual disconnection is correct, then the social contract will slowly come under stress and be weakened (Goodman, 2015). The disconnection between productivity and wages will have many social, economic and political consequences. Inequalities will grow and expectations of a better future for people and their children will lessen. Despair and desperation will grow side by side with apathy (Savage, 2015; Standing, 2014b; McGill, 2016). Thus, the ingredients for a perfect social storm will be brought forward by cascades of innovations that will most likely lead to economic and social crises (Johannessen, 2016). It seems reasonable to assume that it will only take a little spark for the social storm to develop into a social revolution (Petras et al., 2013).

The underlying main driver behind this development is information and communication technology (ICT). Moore's Law states that every 18 to 24 months, computer power is doubled (Thackray, 2015). From 1950 up until today, this development has affected people's lives, organizations' development, nations' wealth creation and the dynamics of globalization (Webster, 2004). All levels of society have been affected by these developments. The rapidity of the pace of change even affects us right down to the level of our thoughts – what we think and believe about our own and our children's future (Pilger, 2016; Savage, 2015). One example of these developments is the recent explosive growth in electric cars, which in the not too distant future could be the only vehicles permitted in traffic; a little further into the future, these electric cars may be self-driven. Thus, such a minor aspect of current technological development may result in our cities being less polluted. Consequently, people may move back to the inner cities, resulting in new re-structuring and re-organization of urban areas; this could even result in the rich living in the city centres while the poor move into the countryside (Wacquant, 2007).

Technological developments will very likely lead to new educational offerings, new jobs requiring different skills, and new functions in institutions and organizations (Pilger, 2016).

Then comes the paradox. Those with relatively long university educations will also find that their jobs are taken over by robots. This could prove to be those who thought they were very secure in their jobs, such as lawyers, doctors, radiologists, journalists, teachers, financial clerks, etc. Any jobs that involve instrumental, linear, repetitive processes that can be predicted by some logical algorithm will eventually be threatened by the "super-intelligent" robot (Pilger, 2016; Savage, 2015; Ford, 2016). The reason for this threat is that the classic relationship between productivity, wages and consumption will collapse (Ford, 2016: xvii). The explanation is that the increased productivity to a lesser extent will go to salaries, but find its way to profits (Piketty, 2014). What we will see is that although productivity increases, salaries will not increase at the same rate (Sprague, 2015).

Globalization

Among the factors that drove forward globalization were China's new economic orientation towards the West around 1988 and the fall of the USSR (1991) that followed that of the Berlin Wall in 1989. This led to more than 1.5 billion people being incorporated in the capitalist economy (Stigliz, 2003, 2007; Swider, 2015; Dickinson, 2016). Another essential prerequisite for globalization was the new ICT and Internet that functioned as an integrative mechanism (Brynjolfsson & Saunders, 2013).

There are two main aspects of globalization. The first is that more and more people are drawn into wealth creation, education and out of extreme poverty (Smil, 2012; Valnazarova & Ydesen, 2016). The second is that some areas are depleted, and poverty becomes entrenched (Rodrik, 2011; Standing, 2014b; Rojecki, 2016). Mason (2015: ix–xxi) describes this development by referring to the country Moldova in Eastern Europe which has remained an economic backwater despite globalization.

One of the results of globalization is obvious – the competition economy as we know it from economics textbooks no longer exists (Roat, 2016). In the present day, it is the major international funds and large international corporations that largely control developments (Charnock & Starosta, 2016; Petras & Veltmeyr, 2001). Figuratively, these operate as feudal vassals setting limits for competition and wage conditions; this new brand of capitalism in the fourth industrial revolution is often referred to as "feudal capitalism". These three factors, financial capital, neo-liberalism and feudal capitalism, are radically transforming the international working class (Rojecki, 2016; Charnock & Starosta, 2016). People are seeking better lives for themselves and their families where there are jobs, and where wages are relatively higher (Olds, 2004; Petras et al., 2013). This is resulting in large-scale global migration (Dickinson, 2016), such as from Africa to Europe, but also migration within regional and national areas, such as from rural to urban areas in Europe and the United States (Varoufakis, 2015; Dickinson, 2016).

Those that govern financial capital and have control over production are interested in minimizing labour costs by transferring them across the globe (Chomsky, 2012; Varoufakis, 2015; Petras et al., 2013). In other words, the relative wages share of value creation is reduced while profits are increased (Piketty, 2016; Ford, 2016; Petras & Veltmeyr, 2011; McGill, 2016). Over time, such a development will not only weaken the working class, but also the middle class (Coates & Morrison, 2016; Gupta et al., 2016), which will lead to tensions and conflicts (Goodman, 2015). One consequence of this may be a rebellion against the elite (Petras & Veltmeyr, 2001, 2011).

A simple inference from prospect theory suggests that this elite will try to maintain their positions, their share of the wealth creation and the benefits they have already achieved (Kahneman & Tversky, 2000). This theory, for which Kahneman and Tversky won the Nobel Prize, states, in brief, that people will spend more energy preserving their positions, income and status, than fighting to achieve these benefits (Adriaenssen & Johannessen, 2016). The rebellion mentioned above may come from either the right or left side of the political spectrum, or take on a completely different form than we can imagine today (Mason, 2015; Petras et al., 2013).

There were good financial results over the globe in China, India, Indonesia, Australia, South Korea, Europe, the US, parts of Africa and large parts of South America (Mason, 2015). However, this economic progress collapsed after the financial crisis (Gaskarth, 2015). The question is whether neoliberalism has outlived its usefulness, and the global economy needs to be re-started using a different economic approach (Wiedemer, 2015; Petras & Veltmeyr, 2013). The new economic system that now manifests itself is a form of feudal capitalism, where large funds and major international corporations downplay competition and focus on stability.

One of the results of the ongoing crisis in the world economy is that economic growth has stagnated. Economic growth will stagnate *for the next fifty years* according to OECD (OECD, 2014). Meanwhile, inequalities are exploding both globally and nationally (Piketty, 2014, 2016; McGill, 2016).

Figure 1.2 shows the social mechanisms which drives the fourth industrial revolution.

Robotization

Skilled work is being increasingly carried out without skilled workers, but by robots (Ford, 2016). The robots will take over both knowledge jobs, such as automated legal services, automated cleaning services and self-checkout machines in shops (Brynjolfsson & Saunders, 2013; Winfield, 2012).

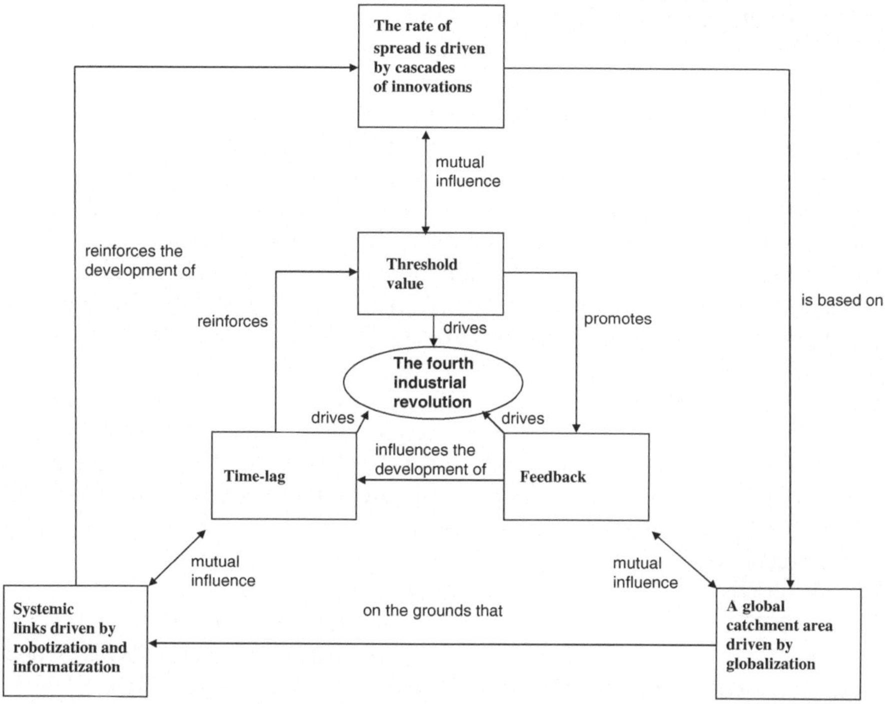

Figure 1.2 Six concepts that summarise the developments of the fourth industrial revolution

The use of robots will change the nature of activities in the following professions: medical, legal, dentistry, teaching, nursing, etc. (Charnock & Starosta, 2016; Valnazarova & Ydesen, 2016). In the not so distant future, robotization of all the service professions and all the knowledge occupations will be an actuality (Coates & Morrison, 2016; Susskind & Susskind, 2015). The main trend is that all the functions that can be digitized will require less human work (Case, 2016).

Odense, a city on the island of Fyn in Denmark is well known internationally by the name of "Robotic Valley". In June 2016) there were 80 companies working with design, construction and assembly of various types of robots. The prognosis for this type of industrial activity is predicted (in this case, Odense) to have an annual growth of 15 per cent. However, it is also reported that there are a lack of skilled workers who can occupy these types of jobs, despite the fact that there is growing unemployment. The types of jobs that are in demand in the robotic cluster in Odense require specialized skills acquired through short, medium and long educations.[1] This would indicate that the fourth industrial revolution's main component, robots, will affect the education market considerably at all levels of competence (Coates & Morrison, 2016; Susskind & Susskind, 2015).

The use of robots in industry is nothing new. For instance, Volvo and other car makes have used them for decades (Winfield, 2012). So what's new? What's new is the fall in the price of robots, their application, the distribution and the use of nanotechnology to increase their functionality (Mataric, 2007). Tesla's factory is an example of how the use of robotization has skyrocketed in auto production. Some of the advantages of robotization are speed, precision, resilience and strength in the execution of work processes (Abd, 2017). Nursing in hospitals and nursing homes, where nursing work is often physically tiring, is one particular domain where increased robotization could make daily work less tiring and more effective. In the near future, robots will also be able to "see" the needs of patients (Xie, 2017), and intervene where necessary. The robots' "eyes" will soon be developed so that they can "see" and understand better than the human eye (Abd, 2017). In the near future, the robots' "eyes" will be able to detect movement, signs, articulations, grimaces and emotional changes (Bleuer et al., 2017). When this happens, many of the jobs in several of the service occupations will disappear or at least be transformed (Susskind & Susskind, 2015).

Robotization will in all probability lead to an explosion of innovation in all, absolutely all, professions, hobbies and functions (Antonelli, 2001; Wilson, 2017). These cascades of innovations will, as we know from innovation theory (Christensen, 2016), destroy established jobs and develop new professions with new competencies (Case, 2016; Johannessen, 2016).

In such a process, the entire education system will change in relation to the requirements of robotization and informatization (Gupta et al., 2016). Such innovations will not only lead to the destruction of established jobs and develop new professions, but will also lead to temporary economic crises (Johannessen, 2016). Out of the crises, large and small, there will emerge a flock of new entrepreneurs (Ford, 2016), who figuratively may be understood as a flock of "phoenix-birds".

Cascades of innovations and the flock of "phoenix-birds" will take off once standards are drawn up for the construction of robots' operating systems (Wilson, 2017). Then entrepreneurs and innovators can concentrate on developing "Lego parts" that fit into an overall design of the individual robots (Zhao et al., 2017).

While innovators with their new inventions create gaps in the market, then entrepreneurs rush in and fill these gaps (Johannessen, 2016). The prerequisite for this development is that there exists a design logic based on the global info-structure, where production, distribution and consumption are integrated into a cohesive system (Lima, 2017). In addition, it will be a condition that standardization creates a modular logic where individual local developers can design and produce their own "Lego parts", which fit into the overall design (Gershuny & Fisher, 2014; Wilson, 2017). Specialized knowledge workers will design and develop "nano" robots which fit into a larger design (Zhao et al., 2017), not unlike some functions in a car engine, or functions in a computer (Brynjolfsson & McAfee, 2014).

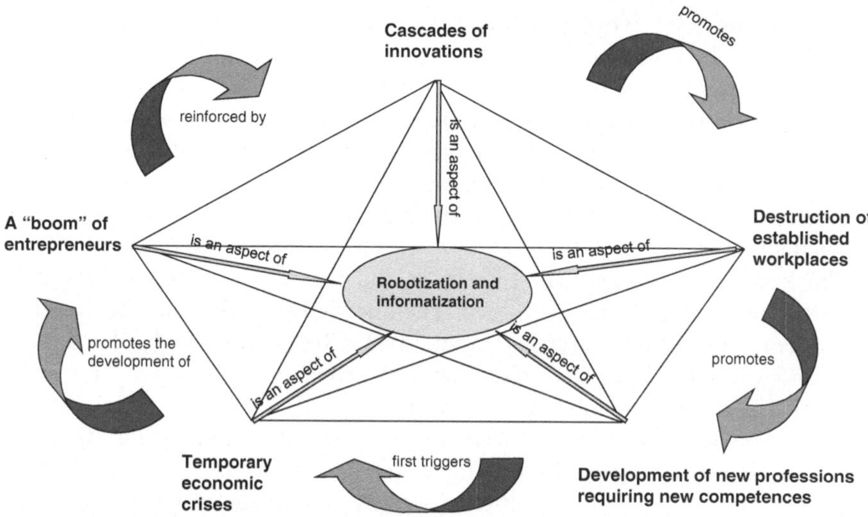

Figure 1.3 Robotization

These new "Lego parts" or functions do not need to be owned by the individual client. The client acquires access to these new functions, for example, via "cloud robotics" in analogy to "cloud-memory" functions that already exist. "Cloud robotics" is data-power in the form of, for example, nano-robot technology that is designed for modular "Lego logic", and is located in the "cloud" and rented by the client. In this way, the costs of robots are lowered considerably because they can be put into mass production. Clients gain access to advanced robotic technology through leases. They can further rent newer levels of computing power through various leases, from free cost, such as the lowest level in a dropbox, to a relatively high rental rate for accessing the desired data-power in various "Lego parts".

What expertise will be in demand in the robotic industry? Design, technology, creativity and innovation will be the winners in this new knowledge explosion (Brockman, 2010: 10–14; Lima, 2017; Xie, 2017). The design, maintenance and modification of robots will require a highly specialized level of expertise (Case, 2016; Zhao, 2017).

Figure 1.3 shows a visualization of the developments that have been analysed and discussed above.

The workplace of the future

Knowledge workers are those who mainly deal with production and the distribution of information (Jemielniak, 2014: 11–22). We can imagine that knowledge work can be divided into two groups. One group are those who are trained for instrumental and linear thinking.

These knowledge workers often have tasks that are repetitive and predictable, for instance, technicians and engineers of various kinds. The second group consists of those who think creatively and systemically, for instance, various types of designers. Designers may work on software design, design of robotic applications, but also the more traditional understanding of design, such as furniture design, fashion design, etc. Designers, working systematically, are expected to create innovative solutions to challenges and problems (Gans, 2016).

The knowledge worker will emerge as the central resource of the fourth industrial revolution (Ford, 2016: xvi). The knowledge worker may be classified into at least four different types (Garza, 2013; Christensen, 2016). Those who have 3–5 years of university education will probably be the new "industrial workers" in the fourth industrial revolution (Ford, 2016; Jemielniak, 2014). Their work will be characterized by instrumental and linear thinking (Catmull, 2014). Many of those with university educations will also take on temporary project positions as self-employed with no relationship to one specific employer. These individuals will be characterized by creative, innovative and systemic thinking (Catmull, 2014). The second main type of knowledge worker will be those with a PhD or equivalent. These may also be divided into two groups. One group, those with T-shaped skills, the knowledge geeks, will be included in the fourth industrial revolution. T-shaped skills means, as the letter T suggests visually, both breadth and depth of knowledge. These knowledge geeks are also characterized by instrumental and linear thinking (Christensen, 2016). Essentially, these individuals will be involved with the maintenance of robots, and they will be highly specialized in programming and robot design.

The second type at the PhD level will be the independent experts. Their mode is creative, innovative and systemic thinking (Catmull, 2014). The knowledge worker in the fourth industrial revolution, in other words, will be a multi-faceted animal, ranging from the traditional "industrial worker" to the independent expert (Susskind & Susskind, 2015).

In Figure 1.4 we have developed a typology of the knowledge workers in the fourth industrial revolution.

The new "hirelings" or precariat go from job to job, and only value their work in so much as it puts bread on the table. They have no fixed relationships to employers but work on short term contracts. This new class, says Standing (2014b: vi–ix), don't relate to 1970s trade union thinking, or to socialist ideology. They rebel against established institutions, comprising both the right and left of the political spectrum (Ross, 2016). The rebellion is aimed upwards towards the elite, irrespective of political leaning (Standing, 2014a). This class is characterized by uncertainty, insecurity, lack of confidence about the future and a perception of being oppressed over a long period (Sennett, 1999, 2013). This oppression has an upper threshold value. When this threshold value is exceeded, the rebellion against their unworthy lives will be ignited. This class of distressed people often show their fury by extremist attitudes and

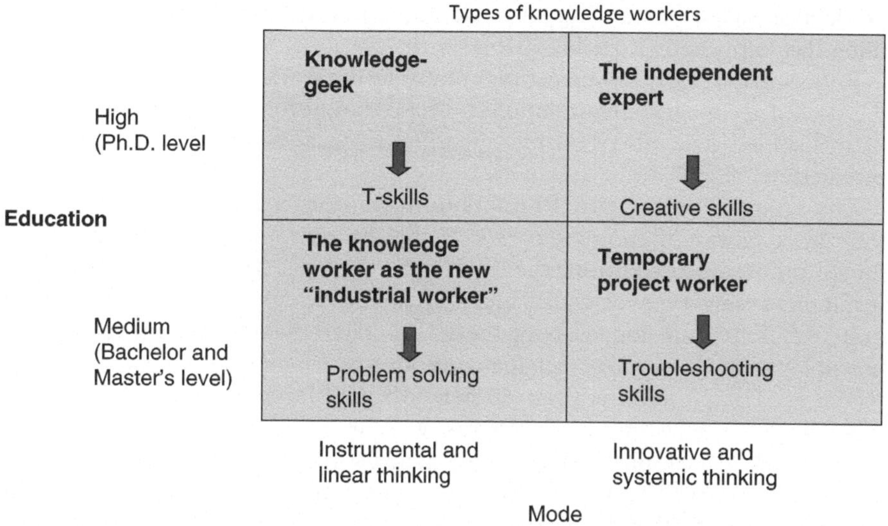

Figure 1.4 A typology of knowledge workers in the fourth industrial revolution

behaviour (Standing, 2014b). Standing (2014b: vii) says: "Across the world, there is an energy building around the precariat." The trend in the industrialized countries (OECD) reinforces the hirelings or the precariat as a class, as more and more are being pushed out of the labour market, or pushed into poorly paid, and meaningless jobs.

Those who feel that their jobs are insecure, and have little or no expectations of a better future, also experience a growing frustration in relation to globalization and robotization (Srinivasa, 2017; Noonan & Nadkarny, 2016; McGill, 2016). This is also occurring in poorly paid service jobs (Ross, 2016).

The 2000s has witnessed a new type of capitalism, a transition from nation-state industrial capitalism, to global capitalism with feudal structures, which may be termed feudal capitalism.

The manufacturing jobs that in the 1980s demanded 2,000 industrial workers, skilled and mostly unskilled, require today 140 skilled workers with special expertise (Ford, 2016: 9). This description is derived from a production facility in California; if it gives a fairly accurate picture generally, then this tells us a lot about what we can expect in the future. Competition for jobs will increase, without necessarily an increase in wages, because there will be so many competing for the jobs that wages will be lowered even further. Profits will grow, and inequalities will increase exponentially (Piketty, 2016). Robotization will result in some of the production be reintroduced to the old industrialized countries from the low-cost countries, but the jobs will not follow. The rationale is that labour will be in less demand, because the robots will have replaced human labour (Monbiot, 2016). Thus, in the period between when the jobs were re-located to low-cost countries in Asia, and the production taken "home" again to domestic industries, i.e. re-shoring, something has happened (Rojecki, 2016).

Robotization leads to at least two things at the workplaces. Firstly, many of the unskilled jobs become redundant, because of automation. Secondly, many of the skilled jobs also become superfluous because robotization increases productivity significantly.

The example Ford (2016: 9) refers to in California leads to a reduction of jobs by a factor of more than 10. If this becomes a characteristic of the fourth industrial revolution, then we have problems. This would most likely result in a new type of mass unemployment, which would be something quite different from the crisis of the 1930s. There will be a great increase in unemployment despite the fact that new jobs will be created and productivity increases. The paradox is that the new jobs will largely be so-called project jobs based on part-time work (Petras & Veltmeyr, 2001, 2011). This development will also apply to highly educated people with university education (Monbiot, 2016). The California story mentioned above could be a forerunner of even further developments such as fully automated manufacturing, which will only need a few highly specialized engineers to maintain the robots. This could lead to the social contract being completely pulverized and feudal capitalism becoming a reality.

This predicted mass unemployment will most likely develop social mechanisms to hide unemployment, such as "citizen salaries" combined with lifelong learning.

Re-shoring will also result in lower transport costs, because the goods can be produced nearer the market. This would be advantageous regarding the reduction of emissions into the environment, but would negatively affect the transport industry.

The other aspect of this development is that automation will also be established in developing countries, such as China, Vietnam, Indonesia and India. With low capital costs, low labour costs and an extremely flexible labour force, which often live in large dormitories outside the factories (Ford, 2016: 12), re-shoring will come to a halt. Competition from developing countries would again threaten jobs in the industrialized world. Thus, after a brief period of re-shoring, the developing countries, by investing in informatization and robotics, will out-compete the jobs in the industrialized countries.

In such a scenario, the OECD area will develop mass unemployment that will likely lead to social conflicts that may threaten the stability of these countries (Petras & Veltmeyr, 2011).

The analysis above is visualized in Figure 1.5.

Conclusion

The problem for discussion in this chapter was: What constitutes the fourth industrial revolution?

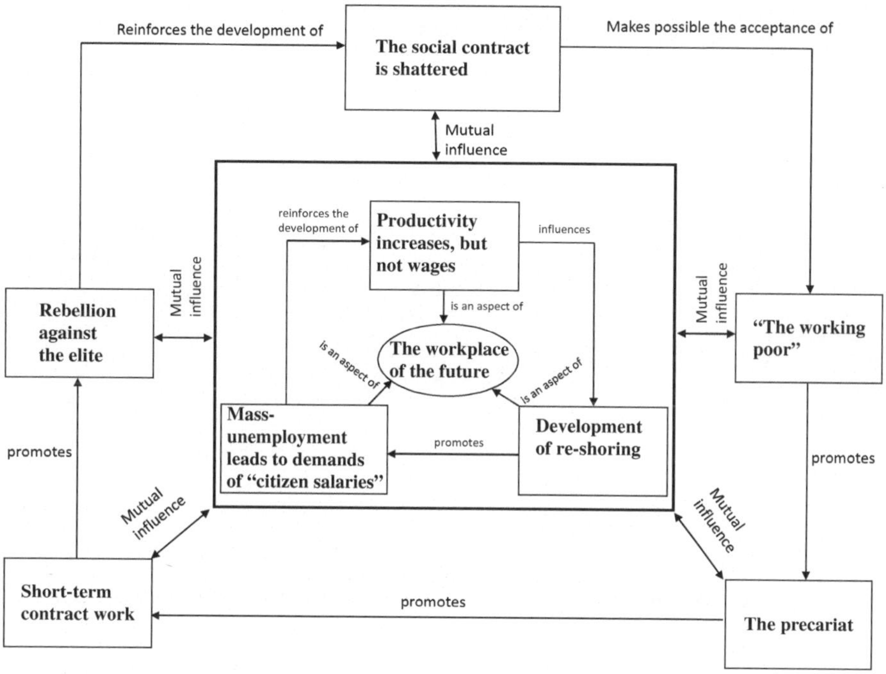

Figure 1.5 Aspects of the future workplace: a systemic overall perspective

In the fourth industrial revolution, thinking will evolve from linear to interactive or disruptive. While the third industrial revolution was about the digitalization of work, in the fourth industrial revolution, robots will bring about the informatization of work. Many of these robots will be systematically connected, such that they can obtain updated information and learn from their own and others' mistakes.

The way we work, where we work, what we work on and our relationships with our colleagues and employers are all in a state of change. The workplace of the future will not necessarily be a fixed geographical location, but may be geographically distributed and functionally divided. In this context, "functionally divided" means that the workplace of the future will largely consist of temporary project-base roles with people performing different functions for different employers. Working hours will also become individualized, i.e., the 9–5 working day will be consigned to history and people will work as if they were self-employed entrepreneurs, although in reality they will be working as subcontractors for various clients. Management in this scenario will be about coordinating individualized, distributed workplaces, and coordinating workers who will largely be self-managing within the framework of fixed contractual targets.

The answer to the problem is presented in Figure 1.6, which is a revision of Figure 1.1.

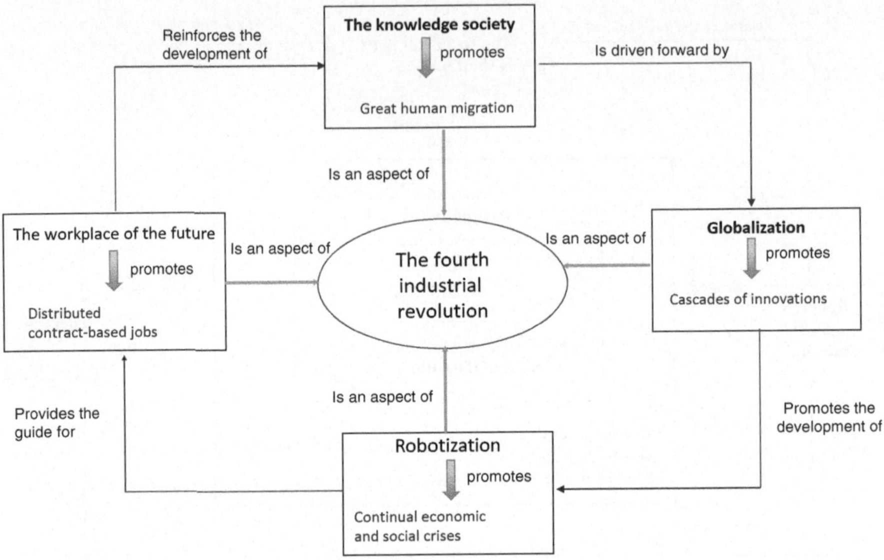

Figure 1.6 What constitutes the fourth revolution, and some assumed consequences

Note

1 *Politikken* (a Danish newspaper), 10th June, 2016, p. 1.

References

Abd, K.K. (2017). *Intelligent scheduling of robotic flexible assembly cells*, Springer, London.

Adriaenssen, D. & Johannessen, J-A. (2016). Prospect theory, *Problems and Perspectives in Management* 14, 2: 84–92.

Antonelli, V. (2001). *The microeconomics of technological change*, Oxford University Press, Oxford.

Barrat, J. (2015). *Our final invention*, St Martin's Griffin, London.

Bauman, Z. (2013). *Does the richness of the few benefit us all?* Polity, London.

Brockman, J. (Ed.) (2010). *This will change everything*, Harper, New York.

Bleuer, H., Bouri, M. & Mandada, F.C. (2017). *New trends in medical and service robots*, Springer, London.

Brynjolfsson, E. & McAfee, A. (2011). *Race against the machine*, Digital Frontier Press, New York.

Brynjolfsson, E. & McAfee, A. (2014). *The second machine age*, W.W. Norton & Company, New York.

Brynjolfsson, E. & Saunders, A. (2013). *Wired for innovation: How information technology is reshaping the economy*, The MIT Press, London.

Case, S. (2016). *The third wave*, Simon & Schuster, New York.

Catmull, E. (2014). *Creativity Inc.*, Bantam Books, New York.

Charnock, G. & Starosta, G. (2016). *The new international division of labour: global transformation and uneven development*, Palgrave, London.

Chomsky, N. (2012). *How the world works*, Hamish Hamilton, London.
Christensen, C.M. (2016). *The Clayton M. Christensen reader*, Harvard Business School Press, Boston.
Coates, K.S. & Morrison, B. (2016). *Dream factories*, Dundum, London.
Dickinson, E. (2016). *Globalization and migration*, Rowman & Littlefield, London.
Ford, M. (2016). *The rise of the robots: technology and the threat of mass unemployment*, Oneworld, London.
Frankish, K. (2014). *The Cambridge handbook of artificial intelligence*, Cambridge University Press, Cambridge,
Gans, J. (2016). *The disruption dilemma*, The MIT Press, Boston.
Garza, D. (2013). *Disrupting*, Create Space, New York.
Gaskarth, J. (ed.). (2015). *China, India and the future of international society*, Rowman & Littlefield, London.
Gershuny, J. & Fisher, K. (2014). *Post-industrious society: why work time will not disappear for our grandchildren*, Center for Time Use Research, Department of Sociology, University of Oxford, Oxford.
Goodman, J. (2015). *Crisis, movement, management: globalising dynamics*, Routledge, London.
Gupta, S., Habjan, J. & Tutek, H. (2016). *Academic labour unemployment and global higher education: neoliberal politics of funding and management*, Palgrave, London.
Jemielniak, D. (2014). *The laws of the knowledge workplace*, Gower, London.
Johannessen, J-A. (2016). *Innovations leads to economic crises: explaining the bubble economy*, Palgrave, London.
Johnson, M. (2014). *Precariat, labour, work and politics*, Routledge, London.
Kahneman, D. & Tversky, A. (2000). Prospect theory: an analysis of decision under risk, in Kahneman, D. & Tversky, A. (eds). *Choices, values and frames*, Cambridge University Press, Cambridge, pp. 17–43.
Lima, P.U. (2017). *Autonomous mobile robotics: a system perspective*, CRS Press, New York.
Mataric, M.J. (2007). *The robotics primer*, MIT Press, Boston.
Mason, P. (2015). *Postcapitalism: a guide to our future*, Allen Lane, London.
McGill, K. (2016). *Global inequality*, University of Toronto Press, Toronto.
Monbiot, G. (2016). *How did we get into this mess*, Verso, London.
Noonan, N.C. & Nadkarny, V. (2016). *Challenge and change*, Palgrave, London.
OECD (2014). *Policy Challenges for the next 50 years*, OECD, Brussels.
Olds, K. (2004). *Globalization and urban change*, Oxford University Press, Oxford.
Smil, V. (2012). *Global catastrophes and trends*, MIT Press, Boston.
Petras, J. & Veltmeyr, H. (2001). *Globalization unmasked*, Zed Books, London.
Petras, J. & Veltmeyr, H. (2011). *Beyond neoliberalism: a word to win*, Routledge, London.
Petras, J., Veltmeyr, H. & Marquez, H. (2013). *Imperialism and capitalism in the twenty-first century: A system of crises*, Routledge, London.
Piketty, T. (2014). *Capital in the twenty-first century*, The Belknap Press of Harvard University Press, Boston.
Piketty, T. (2016). *Chronicles: On our troubled times*, Viking, London
Pilger, J. (2016). *The new rulers of the world*, Verso, London.
Reinert, C.M. and K.S. Rogoff. (2009). *This time it's different: eight centuries of financial folly*, Princeton University Press, Princeton.
Roat, H. (2016). *Capital and collusion: the political logic of global economic development*, Princeton University Press, Princeton.

Rodrik, D. (2011). *The globalization paradox*, Oxford University Press, Oxford.
Rogers, E.M. (1962). *Diffusion of Innovations*, Free Press, Glencoe.
Rojecki, A. (2016). *America and the politics of insecurity*, John Hopkins University Press, New York.
Ross, A. (2016). *The industries of the future*, Simon & Schuster, London.
Savage, M. (2015). *Social class in the 21st century*, Penguin, London.
Schwab, K. (2016). *The fourth industrial revolution*, World Economic Forum, Geneva.
Sennett, R. (1999). *The corrosion of character*, W.W. Norton, New York.
Sennett, R. (2003). *The fall of public man*, Penguin, New York.
Sennett, R. (2013). *Together*, Penguin, New York.
Shipler, D. (2005). *The working poor*, Vintage, New York
Sprague, S. (2015). What can labor productivity tell us about the US economy, *US Bureau of Labor Statistics, Beyond the numbers* 3, 12 (May).
Srinivasa, R. (2017). *Whose global village: rethinking how technology shapes the world*, NYU Press, London.
Standing, G. (2014a). *A precariat charter*, Bloomsbury, London.
Standing, G. (2014b). *The precariat: the new dangerous class*, Bloomsbury Academic, New York.Stigliz, J. (2003). *Globalization and its discontents*, Penguin, New York.
Stigliz, J. (2007). *Making globalization work*, Penguin, New York.
Susskind, R. & Susskind, D. (2015). *The future of professions: how technology will transform the work of human experts*, OUP, Oxford.
Swider, S. (2015). *Building China, informal work and the new precariat*, Ilr Press, London.
Thackray, A. (2015). *Moore's law*, Basic Books, London.
Valnazarova, A. & Ydesen, C. (2016). *UNESCO without borders*, Routledge, London.
Varoufakis, Y. (2015). *The global minotaur*, Zed Books, London.
Wacquant, L. (2007). *Urban outcast*, Polity, London.
Wacquant, L. (2009a). *Prisons of poverty*, University of Minnesota Press, New York.
Wacquant, L. (2009b). *Punishing the poor*, Duke University Press, London.Webster, F. (2004). *The information society reader*, Routledge, London.
Wiedemer, D., Wiedemer, R.A. & Spitzer, C.S. (2015). *Aftershock*, Wiley, London.
Wilson, M. (2017). *Implementation of robot systems*, Butterworth-Heinemann, New York.
Winfield, A. (2012). *Robotics*, OUP, Oxford.
Xie, S. (2017). *Advanced robotics for medical rehabilitation*, Springer, London.
Zhao, J., Feng, Z., Chu, F. & Ma, N. (2017). *Advanced theory of constraint and motion analysis for robot mechanisms*, Academic Press, London.

2 Globalization

The emergence of “Mamounia”, the new global nation

Introduction

As a topic, globalization came to prominence in the 1980s and 1990s. Explanations for the rise of globalization include the fall of the Berlin Wall in 1989 and China’s entry into world markets in the 1980s[1]. Other important factors were the emergence of modern communications technology and the Internet. International power politics became more significant, since there was much at stake when the Cold War ended following the fall of the Berlin Wall and the collapse of the Soviet Union.

New communications technology and the Internet, together with the growth in international trade, have made financial institutions into global operators. The fact that international trade reached approximately 1.5 billion additional people relatively quickly had an effect on economic development worldwide. Trade lifted many people out of poverty, including in China and India. This trend also resulted in some of the people involved in global capitalism becoming extremely wealthy. These people became members of an elite, known as the One Per Cent (Dorling, 2015). As time went on, the members of this new class adopted lifestyles that meant they could be considered a separate nation whose members were more or less exempt from their obligations in the countries in which they lived. The identity of this new One Per Cent class became linked with the notion that it is the rich who create jobs for everyone else, or that benefits and privileges for the wealthy trickle down to others. The Norwegian proverb “When it rains on the priest, the sexton also gets wet” illustrates this “trickle-down economics” (Lehmans, 2015). Such ideas led the One Per Cent and their intellectual brothers-in-arms to believe that they were entitled to certain privileges, e.g. lower taxes (Sowell, 2012).

One of the results of globalization and the emergence of this new One Per Cent was the transfer of production from countries that had become wealthy due to industrialization to parts of the world with significantly lower production costs. As time went on, this caused rich countries to become poorer in relative terms. Accordingly, one could claim that one result of the emergence of the new One Per Cent class has been to make wealthy countries poorer, while increasing the wealth of the One Per Cent to almost unimaginable levels. This trend, whereby a small number of people have become extremely wealthy, while many others have seen their jobs disappear overseas to China, Vietnam, Indonesia, India etc., triggered the emergence of forces opposed to globalization. These anti-globalization forces appeared on both sides of the political spectrum. We saw this in 2016 in the American election, where Donald Trump represented right-wing opponents of globalization, while Bernie Sanders represented their left-wing counterparts. We saw similar trends in Brexit, in Hungary, in Italy, in Austria, in France, etc. The opponents of globalization are on the advance: in some places, they are overturning governments, in others they are threatening the status quo.

Globalization puts the concept of the nation state under pressure (Martell, 2010: 1). In particular, welfare states in Europe have come under pressure. The wave of migration that occurred during the first decade of the 21st century further increased pressure on the economic system that formed the basis of the welfare state. Roughly at the same time as the emergence of globalization, a new economic philosophy developed: neo-liberalism (Eagleton-Pierce, 2016). One of the consequences of this new brand of economics was the deliberate engineering of economic disparity. Other consequences of neo-liberalism included major economic and social problems in many developing countries (Martell, 2010: 6).

Economic, cultural and political globalization, as well as the globalization of social challenges through migration, are increasingly putting pressure on nation states. One explanation for this situation is that it is becoming ever more difficult to solve these problems within the borders of individual nation states. Instead, these problems must be resolved regionally and globally. This makes it necessary to undertake political, economic and social interventions at a supranational level. This has consequences that include a reduction in the influence of democratic institutions. The power to get things done is being transferred to fora where democratic control is for the most part absent (Bauman, 2013; Chomsky, 2012).

This chapter explores the following research problem: How does globalization constitute an aspect of the fourth industrial revolution?

We have formulated three research questions in order to investigate this problem:

1 How does the exercise of power constitute an aspect of the fourth industrial revolution?
2 How does financial globalization constitute an aspect of the fourth industrial revolution?
3 How does the new global "nation" constitute an aspect of the fourth industrial revolution?

Figure 2.1 provides an overview of the introduction. Figure 2.1 also illustrates the structure of this chapter.

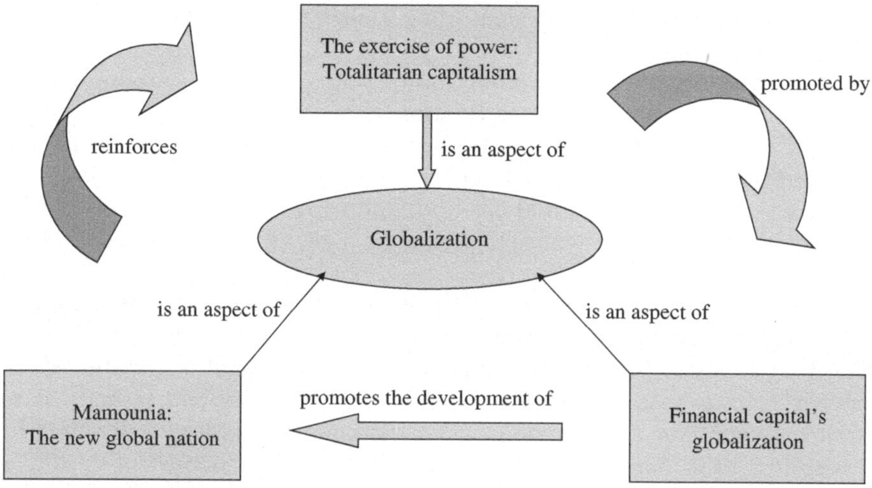

Figure 2.1 Globalization

The exercise of power: totalitarian capitalism

The United States, Russia and China have clear intentions to both threaten and actually use military force in order to achieve dominance in areas they consider to be within their sphere of influence (Chomsky 1999, 2014, 2015b, 2016). This has given rise to a strategy that is explicit in the United States, and practised in China and Russia, of initiating preventive wars for the purposes of securing national interests (Chomsky, 2004, 2015a). In this competitive global environment, international law is mere hot air, and is applied only to protect the superpowers' own interests (Chomsky, 2016). This was the case, for example, with the invasion of Iraq. It is not the case, however, that all nations have the right to conduct this kind of a preventive military strategy. According to Chomsky (2004: 239), the right to undertake such aggressive acts is reserved in general for the United States and its allies. A historian and adviser to the Kennedy administration, wrote that the invasion of Iraq had imposed on the United States: "a global wave of hatred of American arrogance and militarism."[2] The invasion of Iraq did not reduce terrorism; in fact it had quite the opposite effect. The invasion of Iraq created a new militant movement, Islamic State, of which we have subsequently borne the consequences (Cockburn, 2015).

Al Qaeda was also the result of American support for Afghan resistance to the Soviet Union during the 1980s (Burke, 2004). Both Al Qaeda and Islamic State were thus the result of deliberate American policy, even if the consequences that Americans and others ultimately had to bear were not.

Globalization has made military force, and the exercise of it, an active strategy for achieving dominance on the world's markets. This is particularly true of the United States, but Russia and China are following the same path (Raskin, 2015).

On this basis, it seems evident that globalization leads to inequality, insecure employment due to extreme global competition, and competition for lower wages resulting in a reduction of wages relative to profits (Piketty, 2014, 2016). In addition, closer relationships are developing between the business world and politics (Reich, 2009).[3] The incursion of business into politics became all the more evident when billionaire Donald Trump won the 2016 American presidential election, and with the subsequent formation of his new administration consisting of several key players from the world of business.

What then is the psychological "glue" of globalization? One economic psychological mechanism which may be important is the belief in self-interest. The spirit seems to be that what is good for the individual is good for everyone else. However, this idea, on which the capitalist system is built, is an erroneous economic and psychological construction according to Reich in his best-selling book *Supercapitalism* (Reich, 2009: viii). Amongst other things, the construction is faulty because the models it uses become in their consequences the empirical data being examined. The rationale is that if enough people use these models and this thinking, after a period of time one will find support for the models in practice. The models used thus become self-fulfilling prophecies. The faulty construction does not take into account people's sense of justice. When we see that the few are given benefits that the majority do not receive, such as tax reductions for the rich, then we are no longer willing to enter into a social contract that is based on the free market. We realise that the "free" market is not free at all but rigged for the profit of the few (Chomsky, 1999, 2015a, 2015b; Bauman, 2013). "Free" is used as a concept in this context because who wishes to be unfree? However, the free market is not as "free" as some would claim it to be. In other words, the "free" market is not so much an empirical reality but rather an ideological assumption. That is, it is a belief system and not an empirical system. Yet the free market is assumed to be an empirical fact.

According to Reich, one of the consequences of globalization is the decline of democratic capitalism and the rise of totalitarian capitalism (Reich, 2009: x). However, it is not only in China and Russia that totalitarian capitalism has taken hold, it has also manifested itself in the US and EU. Although the façade in the US and the EU may be democratic, in reality the economic system has gained an increasing influence on the political system (Chomsky, 1999, 2004; Freeland, 2013). By totalitarian capitalism, we mean that the economic subsystem has penetrated the political subsystem, dismantling democratic processes. In other words, there is much to suggest that democratic processes are being undermined because the administration of states is hampered by global financial directives.

Paradoxically, although the capitalist system has developed in a more totalitarian direction, consumers have gained greater choice of products, and many people are experiencing increased prosperity since the start of globalization in the 1980s (Reich, 2009: 7).

Managerial salaries have greatly increased since the development of totalitarian capitalism (since the 1980s). In most countries, leaders of medium-sized enterprises often receive higher salaries than the country's political leaders. In itself, this fact sends a clear signal of what is viewed as being important in society: leading a democratic country, or leading a medium-sized private business?

Both Milton Friedman (2002: 10) and Robert Reich (2009: 9), although they stand far apart on the political spectrum, claim that capitalism is a pre-condition of democracy. However democracy is by no means a pre-condition of capitalism. We have witnessed this in China and to some extent in Russia. China has a free market and a capitalist economy but politically is hardly what most people in the West would associate with democracy. We have witnessed the same in Chile under the military junta of 1973, in Spain under the dictator General Franco, in Italy under Mussolini, and in Germany under Hitler. The importance of capitalism as a social mechanism of democracy is most likely exaggerated.

So far, this analysis has attempted to show that capitalist enterprises and democracy are not related elements. They may well exist side by side, but capitalist enterprises can just as well exist side by side with totalitarian governments, fascist governments, military dictatorships and communist regimes. Milton Friedman and Robert Reich's assertion that capitalism is a pre-condition for democracy hardly corresponds with historical reality. However, it is perhaps not difficult to understand why they hold such an ideological assumption. It is based on the belief that capitalism promotes democratic development, which is both right and wrong at the same time.

There are many who claim that the large multinational corporations have taken over in areas where national governments were previously the key players. However, as Robert Reich correctly points out, it is rather the case that the large corporations have less influence after the increase in competition (Reich, 2009: 10). It is rather the large global investment funds that increasingly govern developments. These funds are managed by people who have only one single factor to guide their actions in financial activities: profit. If they ignore this factor, then the members of the pension funds and so on will react through their representatives at general meetings and in other decision-making forums. Thus, from being controlled by the major global corporations a more hidden form of governance is now emerging and is being exercised by those people managing the major investment funds.

The development of global capitalism is not so much controlled by a few large global corporations, but is rather dependent on a way of thinking. This "way of thinking" has many names, such as neo-liberalism, neo-conservatism and so on. However, the core of this ideology is the belief that the market solves all problems. Some of the elements of this ideology are: free trade, deregulation, common standards and privatization, etc. (Harvey, 2007; Petras & Veltmeyr, 2011).

Totalitarian capitalism has not evolved because of a few large corporations controlling developments. Rather, totalitarian capitalism is rooted in a way of thinking, where economic growth, free trade and profit are the principal modes of thought. It is the consequences of such thinking that have led the world into totalitarian capitalism. We see an unvarnished version of it in China and Russia. However, it is also evident in the EU and the United States, through, amongst other things, the invasion of other countries in order to gain control of developments or in the way the European Central Bank and the World Bank treated Greece in 2015 after a democratic election (Varoufakis, 2015).

The road to totalitarian capitalism goes via China, Russia and a form of capitalism in the West that increasingly puts profits in front of people's prosperity (Chomsky, 2014, 2015a, 2015b, 2016).

The democratic aspects of capitalism have come under pressure from several quarters, including global companies (Reich, 2009: 50). How did this happen? The growing competition brought about by globalization has put costs under pressure. Workers have had to compete with workers in other countries where the wages are low. The results may easily be seen: bankruptcies, relocation of jobs, unemployment and an unworthy life for many. In such a situation, people will often support those who say they can help them achieve a worthy life. As well, we have the scapegoat syndrome. In Europe and the US, the scapegoats are the immigrants, refugees and migrants: the "strangers" that Bauman refers to (Bauman, 2013). This is no less the case when Robert Reich in his best-selling book *Supercapitalism* says that the negative criticism of globalization "is mostly nonsense" (2009: 50). Obviously, someone who supports and is supported by the elite would argue that globalization is mainly positive. In elite thinking, capitalism is the opposite of what is described above, namely "a better life for all". Moreover, is this not what the elite have always asserted, even though we witness millions of people experiencing worse conditions in Greece, Spain, the UK, France, the US and other countries. Of course Reich's book is a bestseller! It maintains the myth that capitalism is for the benefit of all, even when millions of people in the author's own country have been shoved into unworthy lives. Robert Reich (2015) has emphatically pointed this out in his book *Saving Capitalism*. In Reich's book, he is joined by two of the heavyweights in economics, Joseph Stiglitz and Paul Krugman. However, their claims are not necessarily correct despite the fact that they are both Nobel Prize winners in economics. It may rather be the case that they are largely influenced by their own ideology.

Totalitarian capitalism is promoted by the quest for profit which is a fundamental aspect of the same capitalism that Reich, Stiglitz and Krugman are defenders of, even though their ideas place them in the centre rather than in alignment with the neo-liberalists.

Financial capital's globalization

Stiglitz, the Nobel laureate in economics, who defends capitalism says: "Globalization today is not working for many of the world's poor" (Stiglitz, 2002: 214). At the same time he says: "We cannot go back on globalization: it is here to stay" (Stiglitz, 2002: 222). In the industrialized world, the traditional industrial workers have suffered huge losses due to globalization. On the other hand, many, especially in developing countries, have benefited; this applies both to a growing middle class and the industrial workers in the developing countries who work in factories that were moved there from the old industrialised countries.

It is primarily trade, the new technology and access to markets with cheap goods that has meant an increase in living standards for millions of people in, for instance, China, India, Vietnam and other countries in Asia (Giddens, 2002; Martell, 2010). The increased prosperity has also led to better health in these countries (Gaskarth, 2015; Swider, 2015). Stiglitz (2002: 214) says that it is not so much globalization that is the problem, but the way globalization has been managed, and the role played by global organizations such as the World Trade Organisation (WTO), the International Monetary Fund (IMF) and the World Bank. However, the problem with Stiglitz's view is that he disconnects the relationship between globalization and globalization's social mechanisms. It is precisely these aforementioned institutions[4] in cooperation with the major multinational corporations and the global investment funds that have fuelled what one calls globalization. Such a disconnection between a phenomenon and that which drives the phenomenon is like disconnecting profits from capitalism, and then saying that capitalism is good, but that it is the profits that destroy capitalism. Without the said institutions, globalization is not globalization, but something quite different. This something quite different is international trade, i.e. international trade driven by internationalization not globalization.

The basis for international trade was bilateral trade agreements. Countries have participated in international trade at least since antiquity, for instance, when Alexander the Great expanded trade to Persia and parts of India. In Roman times, the international trade expanded with the import of pepper and other spices from Southern India (Johannessen, 2016: 9–35). Thus, globalisation and international trade are two different things. Globalization requires a different mindset than internationalization. While internationalization was driven by trade agreements, globalization has been fuelled by free trade agreements, among other things. It is when supranational institutions such as the WTO, IMF and World Bank, as well as multinational corporations and global investment funds, operate outside national frameworks that globalization becomes different from internationalization. We have visualized the differences between trade, trade agreements and free trade agreements in Figure 2.2.

It is when financial capital has woven its network across the global markets that globalization becomes institutionalized (Robé, 2016). With financial capital's globalization, inequality in the industrialized world has greatly increased (Piketty, 2014, 2016). Piketty and others have shown that globalization and free trade have led to increasing economic inequality. Against this background, the IMF and the World Bank have modified their discourse and now focus more on addressing poverty and growing inequality. This is

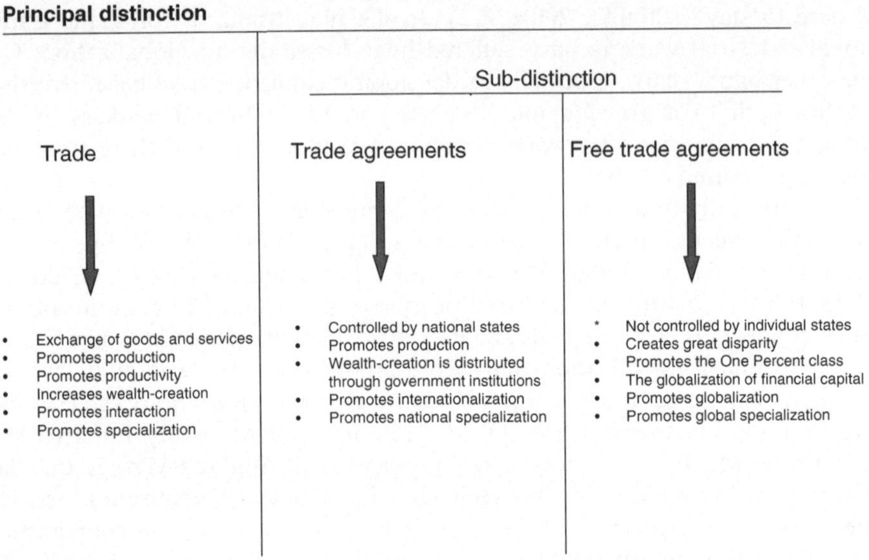

Figure 2.2 The differences between trade, trade agreements and free trade agreements

interesting to note considering that some believe they contributed to the disparities in the first place. Figuratively, this may be likened to a wolf slaughtering some sheep in the flock so the remaining sheep have enough pasture between them to survive.

Market fundamentalism and the globalization of financial institutions are important drivers of globalization (Eagleton-Pierce, 2016). Globalization is based on a fundamentalist ideology (Petras & Veltmeyr, 2011). The problem of globalization is not so much international trade, but rather the fact that it is based on an ideology that promotes the welfare of some at the expense of others (Harvey, 2007). In other words, problems arise when profit is put before people's welfare (Chomsky, 1999; Bauman, 2013). Matters are not helped much when the Nobel laureate in economics, Stiglitz, believes it is the management of globalization that is at fault, rather than globalization as such; and says that "I believe that globalization can be reshaped to realize its potential for good, and I believe that the international economic institutions can be reshaped in ways that will help ensure that this is accomplished" (Stiglitz, 2002: 215).

Financial capital's globalization serves the interests of some, making some people very rich (Piketty, 2014, 2016; Bauman, 2013). However, everybody else has to endure the consequences of the decisions that are made, such as the financial crisis that was triggered in autumn 2007 (Johannessen, 2016: 125–130). Globalization, and especially financial capital's globalization, exerts pressure on democratic institutions in such a way that decisions reached in national democratic forums may be likened to the "decisions" made by a louse in a cod's gills (the louse always has to swim the same way as the fish). In the face-off between national democratic institutions and global finance, there have been few instances in which the will of the former has prevailed.

The financial crisis that was triggered in autumn 2007 is still (in 2018) the cause of major negative consequences for many. The crisis was caused by the pursuit of quick and large profits (Roberts, 2016). Instead of considering the global system as a holistic pattern of interactive individual elements, it was viewed as consisting of separate closed circuits. This has had consequences. When the rate of profit began to fall, the house of cards also collapsed (Roberts, 2016). The negative consequences had to be borne by the many people who ended up unemployed, homeless, socially declassed and social outcasts; moreover, we have witnessed the phenomenon of a disintegrating middle class. The banks and their managements were rescued by the various national governments at the expense of the taxpayers' money (Johannessen, 2016: 125–155).

Regarding the IMF, the World Bank and WTO, they of course genuinely believe that their recommendations serve everyone's interests (Stiglitz, 2002: 216). At least, this was believed until everything started to go wrong around 2007 when the financial crisis erupted. It is often the case that those who talk of "serving everyone's interests" have their own interests in mind, and the interests of those they receive their livelihoods from. It is rarely the case, if ever, that people in private or public institutions speak about their own interests or the interests of their superiors; as a rule, they argue for the interests of everyone. However, it is within their argumentation that we can find the real ideology that lies behind their reasoning. The thinking the said institutions promote is the importance of the market economy, and specifically the American variant (Stiglitz, 2002: 218–222). The benefactors of this kind of market liberalism are the economic elite (Piketty, 2014, 2016; Petras & Veltmeyr, 2011). The international financial institutions follow a logic that is distant from the ideas that people in the West associate with democracy. One aspect of Western democracy is the right of public access to information to enable informed choices. However, the international financial institutions rarely live up to such democratic ideas (Dorling, 2015). Their decisions are taken behind closed doors, and thus consequences are disconnected from their decisions. The democratic institutions lack of access means they are unable to assess the various consequences the closed-room decisions might have for people (Freeland, 2013).

Those who control global financial capital are the masters of the global universe. Global financial capital amounting to billions is moved around in digital space. The power of global financial capital is immense. The rich's capital wealth is handled by the financial elite. A few keystrokes on their part may result in the loss of hundreds of jobs one day and thousands the next (Wacquant, 2009a, 2009b). Over time, they cause the financial tectonic plates to shift resulting in millions of workers becoming unemployed so they need to look for new work. However, many of these unemployed workers do not have the necessary skills to switch to new jobs created in the global knowledge economy. These people may be termed the new "working poor" (Shipler, 2005), or the new precariat (Standing, 2014a, 2014b) in Europe and the US.

Thus, over time, financial capital's keystrokes can change the fate of millions and millions of people. Their keystrokes on their digital screens may be compared to those making a few keystrokes controlling the war machine's drones, bringing death to a small village in Afghanistan or Iraq, an urban district of Aleppo, or a side street somewhere in Libya. The fund brokers on the global stock exchanges in Hong Kong, Singapore, New York, London, Frankfurt, and so on, are the real masters of the global economic combat zone. What these people do collectively can alarm any world leader says Ferguson (2002: 279). Profit is not just *a part* of their activities. *Everything* concerns profit. Their fundamental values revolve around self-interest (Harvey, 2007). The fate of others that is decided by their keystrokes does not constitute a basis for reflection on their part (Chomsky, 1999). Of course, this does not happen because these people are evil. It happens because their actions are guided by a single factor, that of profit (Dorling, 2015; Chomsky, 2016). While self-interest and profits are the gospel of the financial elite (Mason, 2015; Chomsky, 1999; Bauman, 2013), supply and demand constitute their action programme. Their room for reflection is debit and credit. This is what drives them and makes them tick. However, the fact that others have to suffer the consequences of their speculations is not included in their calculations.

The global bond market is enormous. Between 1982 and 1997 it increased by a factor of six to $25 trillion. By 1999, the amount had grown to $34 trillion.[5] This may be understood as the economic foundation of the new nation that figuratively saw the light of day with the rise of globalization. We call this new nation Mamounia. The global bond market is worth more than the GNP of all the world's nations (Ferguson, 2002: 280–281).

Yet, the international flow of capital is nothing new. The Romans used financial capital to extract ore, trade with distant regions and lands, and conduct war (Johannessen, 2016: 9–35). Also in medieval times there were large flows of financial capital between countries. The same was the case in the 1600s and 1700s and later (Kørner, 1995: 507–521; Johannessen, 2016: 35–89). The aim then and now of trade is to make greater profits than is possible within the borders of a single country. There are few people today or in the past who oppose trade as such. However, free trade is a contentious economic topic. Free trade is related to an ideology where neoliberalism is a strong element (Petras & Veltmeyr, 2011). The fundamental idea of neoliberalism is that everyone profits when goods, services, financial capital and labour flow freely between nations (Eagleton-Pierce, 2016). It is claimed beyond doubt that trade promotes economic prosperity (Bonney, 1995). However, it is very doubtful that free trade profits everybody (Harvey, 2007). Micklethwait and Wooldridge (2000) along with several others view globalization in a largely positive light. Others beg to differ, such as the sociologist Giddens (2002) who views globalization as being very problematic for nations, families and individuals. Another sociologist, Sennett, has reached similar conclusions in several of his books (Sennett, 1999, 2003, 2009, 2013).

We already know that some people profit enormously from free trade (Ferguson, 2002: 311; Piketty, 2014, 2016). However, many workers and also those in the middle classes, especially in the old industrial nations, are unable to compete in today's world, because labour costs are lower in countries such as China, India, Vietnam, as well as other countries in Asia (Bauman, 2013; Chomsky, 2015a, 2016). UN figures (1999) showed that three of the world's richest people had greater financial resources than the sum of the gross domestic product of the poorest countries, where 600 million people live (Ferguson, 2002: 311). By 2016, income disparity still had not been reduced (Piketty, 2016).

Global financial capital steers nations towards a unified perception of what is deemed as being the correct economic policies. For instance, recently, this was evident in the case of Greece. Sixteen days after the leftist party Syriza had won the democratic election in Greece in 2015, the European National Bank withdrew their loan from the country (Mason, 2015: ix). It might be said in this context that financial capital clearly expressed its views concerning democratic processes. Greece's Finance Minister, Yanis Varoufakis, had to agree to all the conditions laid down by the European Central Bank, otherwise the country would be staring bankruptcy in the face. The lesson we can draw from the Greek case is simple: politicians do not govern Europe, financial capital does (Mason, 2015: xi). While politicians wish the best for their respective nations, financial capital wishes the best for their investments and profits, and these two goals seldom coincide. Global financial capital does not take into account national interests, but rather the logic of profit. The globalization of financial capital is particularly evident in the case of Greece. Arguably, there was and still is a Greek economic crisis. Varoufakis (2015) makes a pertinent argument when he says that the Greek crisis is only a symptom of a paradigm shift, where there is a transition from an industrial economy to a global knowledge economy. The paradigm shift concerns a transition from national industrial economy thinking, to global knowledge economy thinking.

The dominant global knowledge economy thinking from about 2000 to 2008 was that financial capital was now organized in such a way that risk had been eliminated from financial transactions. However, when risk spiralled out of control into the flaming crisis in 2007–2008 then everyone was greatly astonished. Metaphorically, the fault lay in the fact that they could not see the wood for the trees. They focused on profits and not the system as a whole, with its social, political and cultural consequences.

Financial innovation, like all other types of innovation, is not necessarily a good thing (Johannessen, 2016). In the case of the 2008 crisis, it was definitely a bad thing for most people, but not for everyone. The 2008 crisis, similar to the crisis in 1929, was not triggered because the banks were greedy, although many were. Neither was it triggered by using the wrong economic theory, although parts of the economic theory cannot cope with the run up to an economic crisis. The crisis was triggered due to new financial innovations that created such complexity that no one, absolutely no one, had knowledge of what was happening or the possible consequences.

Innovations lead to economic crises and the financial innovation we saw in the prelude to the economic crisis in 2008 was no exception (Johannessen, 2016: 125–155). One can agree with Varoufakis (2015: 21–22) who compares the capitalist system in a crisis to an injured global minotaur that reacts accordingly when profits are threatened by smashing everything in sight.

One might also say figuratively that the capitalist system is designed like the phoenix bird which sets itself alight and rises from the ashes to start a better life. Economic equilibrium is re-established in the capitalist system when it is able to rise from its own destruction. Thus, capitalism has an in-built social mechanism that drives the social system towards a crisis, but emerges anew from the same crisis. However, there are the millions of people "in the ashes" who never find their way back to the lives they had before. Many of these were tied to industrial production in the OECD area, but there were also millions of the middle classes in the West who were socially de-classed. One might say that the 2008 crisis inflicted severe damage on the middle classes – they just have not fully registered the damage yet because they are still in shock.

Mamounia: the new global nation

According to Ruggie (2016: xii), laws, regulations and business practices have changed dramatically between 2000 and today. We are at what Gladwell (2013) calls a "tipping point", where new power centres, new technologies and the absence of international laws are fostering the development of a new elite, whose members live in a new state, Mamounia. This elite is the One Per Cent, whose members are remote from most other people on the planet (Mason, 2015). In Mamounia, information and communications processes occur between the old continent and the new, which is Mamounia. The old continent consists of low-cost and high-cost countries. Low-cost countries send goods and services to high-cost countries, while high-cost countries invest in low-cost countries, as we have shown in Figure 2.3.

While institutions, laws and regulations are well developed in the old continent, they are absent in Mamounia. A transfer of wealth is occurring from the old continent to Mamounia, i.e., from local states to the One Per Cent who live in Mamounia. The One Per Cent "inhabitants" of Mamounia use the old continent's infrastructure. Most of the "inhabitants" of Mamounia live off the infostructure of the global knowledge economy. This infostructure is disengaged from the control of local states.

Historically, the nations of the old continent were the principal players on the international scene. Together with other states they created the laws and regulations and established the norms and values which were followed by individual states and in the relations between the states. However, in the

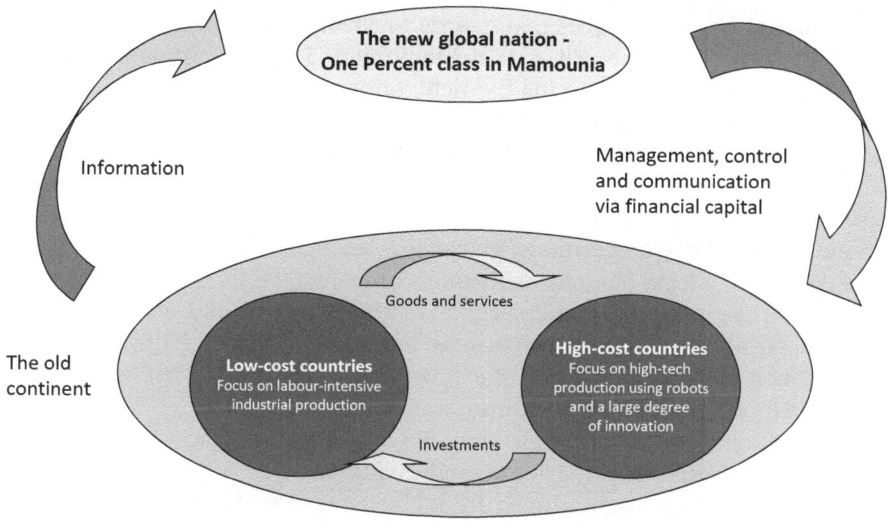

Figure 2.3 Mamounia: the new global nation

global economy things are done differently. First, it was the large multinational corporations that acquired a more central role. This concerned, and to some extent still does, companies such as Exxon, GE, Siemens, Unilever, Sinopec, Google, Coca-Cola, Toyota, Novartis, SAB Miller, and so on. Gradually, the large investment funds also began to acquire greater importance. These include funds such as the Government Pension Fund of Norway, the Kuwait Investment Fund, and so on. These funds have no single company they deliver results to. On the contrary, they convey their results to people such as the hundreds of millions of retirees and future retirees worldwide whose pensions are invested in various funds. Normally, it is only a single guideline that guides their operations and that is the return on invested capital. Of course, guidelines and regulations may also take into consideration ethics and the protection of the environment, and so on. Nonetheless, as a rule, it is profit that is the dominant guideline. In such a situation, financial capital has gained in importance at the expense of both individual states and multinational corporations. It may be said that there are now three players on the international scene: national governments, multinational corporations and large investment funds. This development was made possible through privatization, liberalization, the protection of investors and the privatization of the sovereignty of individual states. By the latter, we mean that the banks and other financial institutions have consciously created opportunities for investors to place their profits in countries with low levels of taxation, and information about the said individuals is virtually absent (Ruggie, 2016: xii). The One Per Cent class have thus been created in the wake of such developments. The One Per Cent class live apart from the 99 per cent; although they may be geographically located in the various states, they live apart in the sense that they are protected by walls, fences and guards to ensure their safety. It is this One Per Cent that inhabit the new nation Mamounia. These are some of the contributing factors that are making the rich nations poorer. The same is the case concerning the multinational corporations. They operate according to a completely different logic than that of state capital. The legal challenges of these multinational corporations are also great, as we have witnessed in the case of Google whose practices have been challenged by the EU's European Commissioner for Competition, Margrethe Vestager.[6] While national states have developed a legal basis for individual companies, "international law barely recognizes the existence of multinational enterprises at all; indeed a number of scholars go so far as to describe it as invisible" (Ruggie, 2016: xiii). It is against such a background that the financial foundation of Mamounia has been developed, as well as its cultural and social foundations.

The multinational corporations are able to greatly influence the policies of national states, especially the terms and conditions that affect the individual corporations. This is partly driven by lobbying, where the corporations have huge financial muscle to influence and bring about decisions in favour of their own views. They can thus exert pressure on national governments to adopt laws that reduce the risk for the corporations, while the individual national states still have to shoulder the risk (Champeil-Desplats, 2016; Chomsky, 2016).

We have reached "the tipping point" (Gladwell, 2013), where the global economy has reached a critical threshold. Gladwell (2013: 7) points out that ideas, information and behaviour spread like viruses do. At the beginning of the global "epidemic", there was only an increasing internationalization of enterprises. The epidemic then spread exponentially like wildfire. In addition, there was the creation of that which is new, that is, an emergent[7] that created a mutation in the virus that was underlying the epidemic. This changed the direction and strength of the epidemic. The results of this epidemic are just beginning to show themselves. Firstly, we have witnessed the emergence of extreme economical inequality, political earthquakes and signs of a cultural revolution in which the political, bureaucratic and intellectual elite have become targets. Secondly, the whole process is making the rich nations poorer, and the poor nations richer. Thirdly, there has been a spike in the fear of "strangers", immigrants, refugees and other migrants. This has created political movements that are not opposed to globalization but are opposed to migration which is one of the consequences of globalization. Fourthly, internationalization has been transformed into globalization. Globalization is qualitatively different from internationalization. Internationalization is based on regular trade agreements, while globalization is increasingly based on free trade and free trade agreements (see Figure 2.2). There is a significant difference between regular trade agreements and free trade agreements. Globalization and free trade agreements create an excess of profits that contribute to building the new nation we call here Mamounia.

Mamounia is an emergent, something that is qualitatively new that has not existed in the world before. The power system of Mamounia is not anchored in any single country. The large multinational corporations and large investment funds play a crucial role within this new power system. The development of the new power system has occurred without any overall plan, but resembles an epidemic in which no one knows who will be affected, when or where. Mamounia is also an emergent in the sense that there are no institutions, citizens, budgets, military system or anything we normally associate with the establishment of a new nation. Thus, Mamounia is in this sense a fiction, but it is also very real for two reasons. Firstly, Mamounia is real in the impact it has on national governments, enterprises and individuals. Secondly, Mamounia is real because the power Mamounia possesses has major consequences for national states. Yet, researchers have no analytical tools with which they can analyse the power system of Mamounia. Researchers have to rely on statistics related to the financial system's development, integration, and the management and control of financial capital. Yet, what we see happening may be described as the development of a new "world power system" (Robé, 2016: 11). Pictorially, the evolution of this power system resembles an octopus whose tentacles are reaching out across the globe. The challenge for researchers is to conceptualise this power system says Robé (2016: 11). This is the task we attempt here, and which is indicatively shown in Figure 2.3 and Figure 2.4.

Mamounia, and the new power system has no constitution, no geographical locality, no citizens in the classical sense and no laws. Mamounia self-institutionalizes its power structures in favour of some and to the disadvantage of others. Like a parasite, Mamounia uses its host to survive. In this way, Mamounia makes the rich nations poorer, by taking without giving. If this maladaptive parasite goes unchecked, it may end up killing its host. If the social system does not expire due to this parasite, it will nevertheless impoverish large groups in the former industrial states, who will end up de-classed, unemployed and living in fear of what the future may bring.

The power relations are developed through self-reinforcing loops. One example is the transfer of labour-intensive work to low-cost countries (Figure 2.2) due to competition focusing on labour costs. This has affected employment in the rust belt in the US, the automotive industry in Detroit, US, and in the industrial areas in Wales, and so on. Jobs disappear and are not replaced by new employment opportunities because the low-cost countries can always compete on labour costs. Millions of jobs have disappeared in this way from the United States (Wacquant, 2007, 2009a, 2009b). The power systems in Mamounia benefit from the competition for low labour costs, because it has an epidemic-like ripple effect. The new jobs that are created have completely different requirements regarding skills, talent, expertise and motivation than the old labour-intensive jobs that disappeared. This also creates a reserve labour force that presses labour costs down in the event of a demand for labour in the future. This class of people goes under various names such as "the working poor" (Shipler, 2005) and "the precariat" (Standing, 2014a, 2014b). Another development is that the gap between the One Per Cent class and the 99% is continually widening, thus increasing the power system of Mamounia.

The new nation we call Mamounia is thus an attempt to conceptualize globalization in a new way. What we do know is that when enterprises in the global space interact with each other and with various types of global markets, an emergent occurs, i.e. something new in the world that did not exist before. This emergent cannot be analysed by dividing it up into its constituent elements. Figuratively this would be like dissecting a bird in order to study its flight characteristics. If one divides the emergent and then attempts to put the pieces back together again into a working whole, it would be like trying to create the proud oak from the sawdust remaining on the patch where the tree was felled. Global structures cannot be analysed as if they existed under the umbrella of one or more states, precisely because it is the emergent system that should be examined. In the global economy, there are per definition no external factors, apart from the weather, the stars and the influence of the sun. The system of power in Mamounia is not influenced by anything other than its own modus operandi.

The units or elements in this power system are the global enterprises that manufacture, distribute and reintegrate following a modular logic. This modular logic consists of an extreme focus on low-costs, high quality, a high level of expertise and a high degree of innovation. This in turn reinforces the development of the global enterprises.

Although there are no institutions, constitution or laws in Mamounia, the power system in Mamounia uses the institutions and laws, such as the contract laws in the countries where Mamounia's "citizens" find the most suitable conditions (Champeil-Desplats, 2016: 159–169).

We have conceptualized Mamounia and the global power system in Figure 2.4.

Conclusion

In this chapter, we have investigated the following issue: How does globalization constitute an aspect of the fourth industrial revolution?

The answer we have arrived at can be understood along three axes. First, there is much to suggest that the exercise of power that we see in the global economy is revealing the emergence of totalitarian capitalism. This is already apparent in, for example, China and Russia. The second axis gives strength to the idea that a totalitarian form of capitalism is the consequence of the globalization of financial capital. Financial capital seems to be the backbone of totalitarian capitalism. At the same time, both financial capital and totalitarian capitalism are fostering the development of a new global "nation", which in this chapter we call Mamounia. Mamounia does not operate in accordance with geographical borders, but with symbolic borders. Mamounia also has no laws, institutions of government or constitution, unlike traditional nation states. Mamounia does however have "citizens". Most of these are the One Per Cent, who obtain information from the old continent, which consists of nation states, and who control these nation states through, among other things, financial capital.

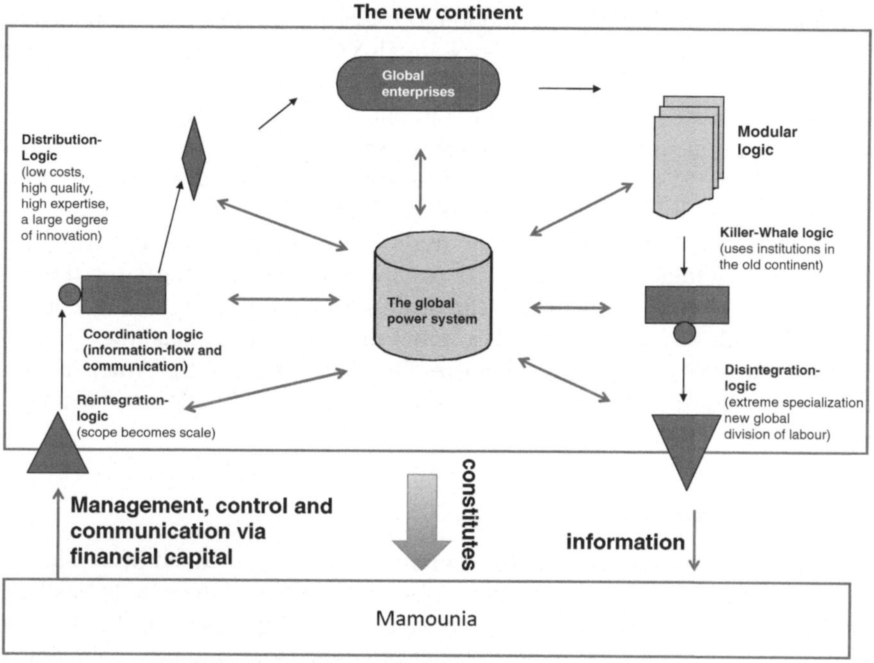

Figure 2.4 Mamounia and the global power system

Theoretical implications

When people feel they have been violated over a period of time, when Mamounia shows its claws and people become afraid of their own future and the future of their children, when their dignity is threatened and they experience a lack of respect, then they will most probably react. Their reaction may take many forms. First, there may be apathy; figuratively, this may be likened to the calm before the storm – even the birds are silent before a storm. However, like a storm, like thunder, like revenge led by the rhetoric of action, the reaction will come. Then figures representing the violated will occupy positions we have only read about from the great revolutions of the world. However, this revolution will take on completely different forms than that of the major revolutions of the past. The reaction will dress itself in new appearances, but the revolution will be similar to previous revolutions, such as the Industrial Revolution, in the sense that it will transform everything that existed before it. The reaction will probably emerge as a nostalgic yearning for the past, the security one felt before globalization and Mamounia ruled the world. The nostalgia will be felt as something good and safe, although objective analyses show that conditions before globalization were far from being safe and secure. Yet, the situation is different for many now, especially those who worked in companies that were re-located to maximize profits.

In the times we live in, the economic crises are more systemic than previously, i.e., they are inter-linked. They are also systemic because many systems are connected at different levels.

This means that reactions to a crisis in one location will rapidly spread to a different location in the global economy. Since the crisis that was triggered around 2007/2008, the so-called financial crisis, the consequences have been felt around the globe. After the fall of the Berlin Wall and China's entry to the capitalist market, conditions for trade have changed. About two billion new people entered into global trade resulting in new challenges for the free market. One of these challenges is the disillusionment of many of the working and middle classes in both the US and Europe. Their anticipatory anxiety about the future has led to discontent resulting in changes in the American and European political landscapes. This dissatisfaction has led people to Nostalgia, the "yearning for the lost country".

Practical implications

The free trade and the free market economy that are embedded in neo-liberal ideology have established Mamounia as a reality. One of the results has been a redistribution of wealth from the working and middle classes in the US and Europe to the working and middle classes of China, Vietnam and India. Statistically, more people in the world are now better off with a global market economy. Nevertheless, this statistical fact is of little consolation to the steel belt workers in the US and the working and middle classes in Wales who have lost their jobs. They have become frustrated, angry, discontented and react accordingly. It is an emotional revolution, not a rational one, we are witnessing. This revolution is taking place through democratic institutions. The workers' nostalgic dream is that they want to return to their jobs in the automotive industry in Detroit, the steel industry in Michigan, the bearings industry in Illinois, the slaughterhouse industry in Denmark, and so on. But unskilled jobs will not be moved back to these regions. Another important point is that the skilled jobs will also follow, if the profits are greater where enterprises choose to re-locate.

Robots will take over in production processes to an increasing extent. Many people had hoped that this would result in jobs moving back to Europe and America, because robots do not require costly labour. In this context, it is worth noting the experience of a 31-year-old robot programmer, Michael Robinson: "I was told that although there will be increasing automation, there will always be a need for people to program robots. Yet, this is not the case as the factory has been moved to Mexico. It's hopeless!"[8] When five million factory jobs have disappeared from the US since the turn of the millennium,[9] then it is understandable that people are desperate and react accordingly. The jobs are not disappearing just because the costs are lower elsewhere; they are disappearing because the profits are higher, which is something quite different. The comments of the robot programmer, Michael Robinson, show this.

Further research

On the other hand, robotization could lead to a return of factories from low-cost countries to the old industrialized nations. The rationale is that labour costs will be of less importance. Robotization will mean that US and Europe will be able to compete on productivity with countries such as China. As such, the "emotional revolution" we are seeing now will have rational positive consequences for the skilled working class.

In this scenario, three elements working together will reinforce the redistribution of wealth we have seen develop since the fall of the Berlin Wall and up until the present. The first element is the global power system. The second element is the global financial system. This system is based on a network logic where the nodes can be found in a few mega-cities such as Hong Kong, London, Frankfurt, Paris, New York, and so on. The third element is what may be termed a transition from the power of national states to globalism and Mamounia. Globalism and Mamounia are also the target of the "emotional revolution". Yet it is highly likely that the rebellion is just a ripple along the way towards increased globalism, because the strongest forces are the thinking related to a free market economy. This will only reinforce globalism and "militarize" Mamounia so that a totalitarian capitalism will spread worldwide.

The explanation for this development is simple. The free market economy with its logic of free trade will reinforce globalism. One of the consequences of the interaction between the global power system, the global financial system and globalism is that the redistribution of wealth will be reinforced. In its consequences, we will see that the rich nations become poorer because wealth creation will be moved to where the rate of profit is highest. In turn, this development will reinforce the "emotional revolution".

Political psychology seems now to be relevant in relation to globalization. People who in principle supported competition and trade now may find themselves without a job, socially declassed or threatened by both of these phenomena. In this context, one might say they are suffering the consequences of their views, or that the chickens have come home to roost.

We conclude this chapter by referring to the paradox of the philosopher Zappfe: It's what you are good at that will be your downfall; or, as in this case, capitalism is so good at competition that it will lead to its downfall.

Notes

1 In 1988 in a speech at a Communist Party conference, the Chinese President Deng Xiao Ping said that China would become rich through exports and allowing foreign capital to invest in China.
2 Arthur Schlesinger in the *Los Angeles Times*, 23 March 2003.
3 Robert Reich is Professor of Public Policy at the University of California, Berkeley. He has served in the Carter, Ford and Clinton administrations. (https://no.wikipedia.org/wiki/Robert_Reich)
4 WTO, IMF, the World Bank.
5 Financial Times, 4 February 1998, ref. in Ferguson, 2002: 279 and 468.

6 http://time.com/3890416/vestager-google-antitrust/
7 An emergent occurs if something new turns up on one level that has not previously existed on the level below. With emergent, we mean: "Let S be a system with composition A, i.e. the various components in addition to the way they are composed. If P is a property of S, P is emergent with regard to A, if and only if no components in A possess P; otherwise P is to be regarded as a resulting property with regards to A" (Bunge, 1977: 97).
8 Article in the Norwegian newspaper, *Aftenposten*, 5th December 2016.
9 *Aftenposten* 5th December 2016.

References

Bauman, Z. (2013). *Does the richness of the few benefit us all?* Polity, London.
Bonney, R. (1995). *Economic systems and state finance*, Oxford University Press, Oxford.
Bunge, M. (1977). *Treatise on basic philosophy. Vol. 3. Ontology I: The furniture of the world*, D. Reidel, Dordrecht, Holland.
Burke, J. (2004). *Al Qaeda*, I.B. Tauris, New York.
Champeil-Desplats, V. (2016). Constitutionalization outside of the state? in Robé, J.P., Lyon-Caen, A. & Vernac, S. (eds.) *Multinationals and the constitutionalization of the world power system*, Routledge, London, pp. 159–169.
Chomsky, N. (1999). *Profit over People*, Seven Stories Press, New York.
Chomsky, N. (2004). *Hegemony and Survival*, Penguin, London.
Chomsky, N. (2012). *How the World Works*, Hamish Hamilton, New York.
Chomsky, N. (2014). *Year 501: the conquest continues*, Haymarket Books, Chicago.
Chomsky, N. (2015a). *Masters of mankind*, Penguin, London.
Chomsky, N. (2015b). *The Washington connection and third world fascism*, Pluto Press, New York.
Chomsky, N. (2016). *Who rules the world*, Hamish Hamilton, London.
Cockburn, P. (2015). *The Rise of Islamic State*, Verso, London.
Dorling, D. (2015). *Inequality and the 1%*, Verso, London.
Eagleton-Pierce, M. (2016). *Neoliberalism: the key concepts*, Routledge, London.
Ferguson, N. (2002). *The cash nexus*, Penguin, London.
Freeland, C. (2013). *Plutocrats: the rise of the new superrich*, Penguin, London.
Friedman, M. (2002). *Capitalism and freedom*, University of Chicago Press, Chicago.
Gaskarth, J. (ed.). (2015). *China, India and the future of international society*, Rowman & Littlefield, London.
Giddens, A. (2002). *Runaway world: how globalization is reshaping our lives*, Profile Books, London.
Gladwell, M. (2013). *The tipping point*, Abacus, New York.
Harvey, D. (2007). *A brief history of neoliberalism*, Oxford University Press, Oxford.
Johannessen, J-A. (2016). *Innovations leads to economic crises: explaining the bubble economy*, Palgrave, London.
Kørner, M. (1995). Public credit, in Bonney, R. *Economic systems and state finance*, Oxford University Press, Oxford, pp. 507–538.
Lehmans, D.J. (2015). *Understanding trickle down economics*, Create Space, New York.
Mason, P. (2015). Foreword, in Varoufakis, Y. *The global minotaur*, Zed Books, London, pp. ix–xii.
Martell, L. (2010). *The sociology of globalization*, Polity, London.

Micklethwait, J. & Wooldridge, A. (2000). *A future perfect: the challenge and hidden promise of globalization*, Times Books, New York.
Petras, J. & Veltmeyr, H. (2011). *Beyond neoliberalism: a word to win*, Routledge, London.
Piketty, T. (2014). *Capital in the twenty-first century*, The Belknap Press of Harvard University Press, Boston.
Piketty, T. (2016). *Chronicles: on our troubled times*, Viking, London.
Raskin, M. (2015). Foreword, in Chomsky, N. *Masters of mankind*, Penguin, London, pp. 9–19.
Reich, R. (2009). *Supercapitalism*, Icon Books, New York.
Reich, R. (2015). *Saving Capitalism*, Alfred. A. Knopf, New York.
Robé, J.P. (2016). Globalization and constitutionalization of the world power system, in Robé, J.P., Lyon-Caen, A. & Vernac, S. (eds.) *Multinationals and the constitutionalization of the world power system*, Routledge, London, pp. 11–52.
Roberts, M. (2016). *The long depression*, Haymarket Books, London.
Ruggie, J.G. (2016). Foreword, in Robé, J.P., Lyon-Caen, A. & Vernac, S. (eds.) *Multinationals and the constitutionalization of the world power system*, Routledge, London, pp. xii–xvii.
Sennett, R. (1999). *The corrosion of character*, W.W. Norton, New York.
Sennett, R. (2003). *The fall of public man*, Penguin, New York.
Sennett, R. (2009). *The craftsman*, Penguin, New York.
Sennett, R. (2013). *Together*, Penguin, New York.
Shipler, D. (2005). *The working poor*, Vintage, New York.
Sowell, T. (2012). *Trickle down theory and tax cuts for the rich*, Hoover Institution Press, New York.Standing, G. (2014a). *A precariat charter*, Bloomsbury, London.
Standing, G. (2014b). *The precariat: the new dangerous class*, Bloomsbury Academic, New York.Stiglitz, P. (2002). *Globalization and its discontents*, Penguin, London.
Swider, S. (2015). *Building China, informal work and the new*, ILR Press, London.
Varoufakis, Y. (2015). *The global minotaur*, Zed Books, London.
Wacquant, L. (2007). *Urban outcast*, Polity, London.
Wacquant, L. (2009a). *Prisons of poverty*, University of Minnesota Press, New York.
Wacquant, L. (2009b). *Punishing the poor*, Duke University Press, London.

3 Robots and informats will cause economic and social crises

Introduction

In his essay *Economic Possibilities for Our Grandchildren* (1928), John Maynard Keynes predicted that by 2028, living standards in the United States and Europe would be at such a level that no one would need to think about earning money (Keynes, 1963: 358–373). The point he was trying to make was that 100 years into the future, society would have reached such an advanced level that only a few workers would be needed to keep the machines running.

In this context, the word "robot" is used to refer to a computer that is programmed to assist humans in the performance of physical and mental labour. The word robot comes from the Czech word "robota", which means "work" (Abd, 2017). The development of artificial intelligence means that robots are becoming ever more capable of performing mental labour. By informat, we mean robots with artificial intelligence which are interconnected in a global technological network. Figuratively, informats may be imagined as clusters of neurons in the human brain, which are connected to other neuron-clusters to create the various functions of human intelligence (Wilson, 2017; Winfield, 2012; Vadakkepat & Goswami, 2018). In the financial world, the use of robots by financial analysts provides an example of the use of so-called informats. Informats for use in medical surgery are already on the drawing board and will be a reality in the near future (Bleuer et al., 2017); the same applies to the use of informats in the service and education sectors (Bleuer & Bouri, 2017). Informats are understood here in the context of the above description as being emergent in relation to robots; informats can sense, analyse and reach decisions in the space of a micro-second.

In the future, if robots and informats perform most of the heavy physical labour, and mental and intellectual work, respectively, what will be left for people to do? It may be possible that people can then focus on what they do best: think, reflect and communicate. People could then use their time and resources to create a world that many only dream about today. However, this future scenario is not pre-destined. The fourth industrial revolution may easily lead to a situation where only a few profit from the new technology, while the majority will be exposed to economic and social crises. The point being made here is that it is not the robots and informats that will decide how the future unfolds; this will rather depend on what policies people adopt to deal with these technological changes.

It seems that innovation has come to represent the new economic and political "religion". All the social systems are demanding technological, economic and institutional innovations (Barrat, 2015). Of necessity, minor innovations lead to minor social and economic crises, while major innovations lead to major social and economic crises (Johannessen, 2016). In this context, Christensen uses the word "disruptions" when referring to major changes, i.e. an innovation that disrupts and makes existing business models obsolete (Christensen, 2010, 2016). Among its consequences, this leads to economic and social crises, because the old is demolished, while new ways of producing, distributing and consuming emerge.

Robotization with its innovations and disruptions will affect the way we work, employment structure and leisure time. Not least, the relationship between leisure and work will change radically, because robotization will lead to increased production (Abd, 2017).

This article examines the question: How does robotization constitute an aspect of the fourth industrial revolution?

In order to answer this question, we have formulated two research questions:

R1: How do innovation, herd behaviour and economic bubbles constitute an aspect of the fourth industrial revolution?
R2: How does destructive creation constitute an aspect of the fourth industrial revolution?

Figure 3.1 shows a representation of what is described in the introduction; it also shows how the rest of the chapter is organized.

Robots and informats: economic and social crises

Robotization will occur in parallel with social systems developing infostructure and informatization. Infostructure, as it is used here, refers to all the activities and processes that promote informatization in social systems. Miller (1978) refers to ten critical processes that exist on all system levels. These information processes are: 1) information control, 2) information channels and networks for communication, 3) information gathering, 4) information analysis, 5) information strategy, 6) information structuring and systematization, 7) information coordination, 8) information storage and retrieval, 9) information culture, and 10) distribution of information. Informatization here refers to the digitalization of the ten infostructure processes referred to above.

The robotization and informatization of society will occur through extensive expansion of the global infostructure (Brynjolfsson, & McAfee, 2014; Ford, 2016). Robots and informats will compel new ways of thinking and working, because every level of the social system will be affected (Vadakkepat &

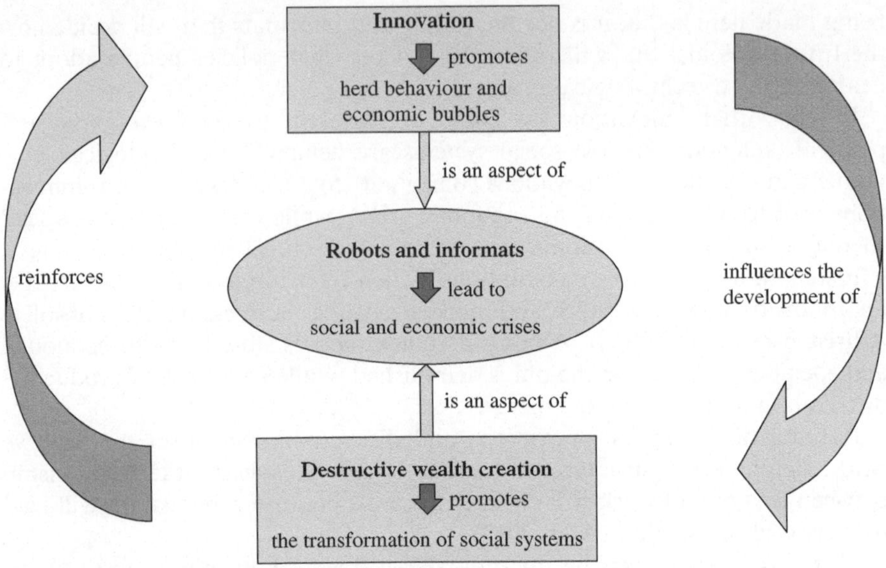

Figure 3.1 Robotization

Goswami, 2018). This will apply to economics, politics, culture and social relationships (Reich, 2015; Frankish, 2014; Gans, 2016). Keynes' essay, mentioned above, is relevant in this context; Keynes says that in such a situation, the greatest challenge people will have to face is related to the question: How will people deal with the freedom that will be created by robots and informats?[1] It is in such a situation that new ways of thinking will become central. The question "How will we live when our freedom from paid employment becomes a reality?" will be a problem for the field of philosophy. How will we spend our leisure time when we do not need to spend as much time making money to live on?

Regardless of the changes to working life brought about by the introduction of robots and informats, everyone, absolutely everyone, will be affected in one way or another (Lima, 2017; Bleuer & Bouri, 2017).

As such, Keynes was prescient in his prediction of future exponential technological development. However, in this context, a pertinent question is: Are we willing to accept a two-level society, in which the 1% class rule through finance capital, and the remaining 99% struggle to make ends meet (Gershuny & Fisher, 2014; Dorling, 2015)? Among the 99%, there will be major differences in terms of employment, income, leisure time and how people are employed.

What we do know with relative certainty is that the more innovations that are introduced into the market, the greater the likelihood of disruption and economic crises (Johannessen, 2016). Robots, informats, the infostructure and the informatization of society will influence how people work, their leisure time and the structure of employment. The informatization of the majority of functions will change the ways in which we work (Lima, 2017; Vadakkepat & Goswami, 2018).

One example of exploitation of informatization and the global information structure for innovative business development is the mobile photo-sharing company Instagram. When Instagram was sold to Facebook for about $1 billion, they had only 13 employees, although they had 30 million customers. On the other hand, Kodak (in a related business sector to Instagram) with 145,000 employees, went bankrupt a few months before the sale of Instagram (Reich, 2015: 206–207). The business models of the two companies were completely different. It is such a total change to business models to which Christensen refers when he uses the term disruption (Christensen, 2010, 2016). Disruption, small and large financial crises, changes in working methods and the use of informats and infostructure to produce, distribute and consume, will all inevitably lead to new ways of thinking. Instagram, Facebook, LinkedIn, and so on, are the results of new ways of thinking. Innovation has become the new economic and political religion.

In the robotized society of the future with its global information structure, if you have an idea you can then sell "the product" from your home, whether this happens to be in Wales or Detroit, to millions of potential customers without having a single employee. The ideas and access to technological expertise will, in such an economy, be the most important social mechanisms for wealth creation. In such an economy, work activities will be divided into various components; one can imagine this consisting of four components.

Firstly, there are those who have ideas and sell them in a global market. Secondly, there is the expert who is concerned with symbol analysis. This could be the financial analyst, the surgeon, the engineer, the psychologist, and so on. Thirdly, there are those who are concerned with routine work, such as the traditional industrial workers, service workers in hospitals and hotels, and bureaucrats in intermediate positions in private and public sectors. Fourthly, there is the professional group whom one could call the "front-line workers". These people are close to those they are providing service to. This includes employees working in nursing homes, home-helpers, teachers, educators, pre-school teachers and all who use their expertise to provide a service or services where physical proximity is important.

We have developed a typology of the four occupational categories that we propose will exist in the fourth industrial revolution (see Figure 3.2).

If we consider those professions under the category of "symbol analysis", such as financial analysts and surgeons, it is highly likely that the nature of their work will change when robots and informats become even more integrated in their work activities. Robots have already been introduced into the field of financial analysis, and in the future will further modify the nature of this area of work, so fewer people will be needed in this type of profession. In the relatively near future, robots and informats will also be able to perform the most complex surgical procedures (Bleuer & Bouri, 2017). Moreover,

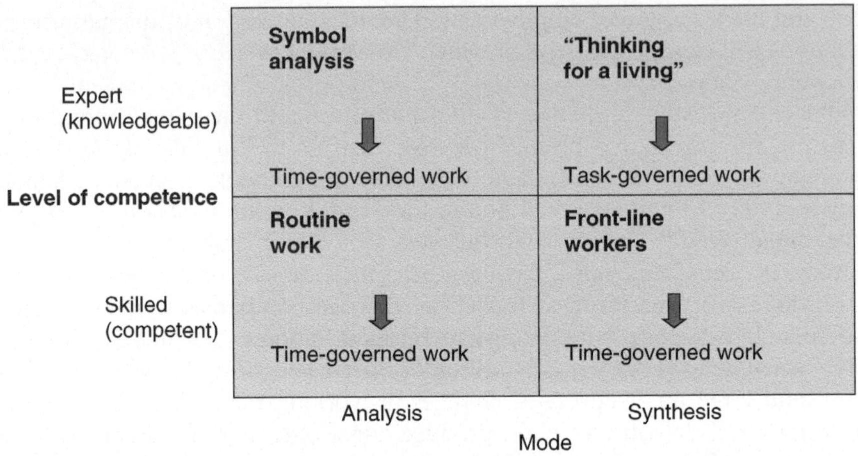

Figure 3.2 A typology of working life in the fourth industrial revolution

informatization and the infostructure will enable the interconnectedness of robots, informats and surgeons. This may result in a situation where the physical presence of the surgeon will not be needed during an operation; for instance, the surgeon may be located in Los Angeles, California, while the patient is operated on in Mehamn near the North Cape in Norway (Bleuer et al., 2017). Such a development would also totally change the way we think concerning the construction of hospitals. If robots, informats and the application of expertise at a distance become significant elements in the future treatment of patients, will the construction of giant hospital complexes still be necessary? In the future, robots, informats, the infostructure and informatization will probably result in the construction of small compact hospitals located, for instance, on street corners, like architectural gems in the centres of small towns. These small compact hospitals will most likely specialize in a particular field of medical care, and be fully equipped with up-to-date technology and robots. These small specialized robotic and informat hospitals will be staffed with service personnel with nursing educations, one or two doctors, and a robotic analysis and surgical unit. Obviously, this is far into the future, but not so far that we cannot already see such tendencies in related organization of technology (Abd, 2017; Bleuer et al., 2017; Bleuer & Bouri, 2017; Vadakkepat & Goswami, 2018).

Many of the jobs under the category of "routine work" shown in Figure 3.2 will, most probably, be moved to countries where costs are low, skills high, product quality good and where there is also a certain degree of creativity that leads to innovation (Brockbank, 2013; Bruce & Crook, 2015). On the other hand, when robots and informats take over the production of tasks under this category, it will not be necessary to move production to low-cost countries. A foretaste of this trend may be found in a report in the McKinsey Quarterly (2016), where it is stated that 78 per cent of routine physical activities in US workplaces will in the future be replaced by technological automation.[2] According to the report, the most susceptible areas of work to automation will be predictable physical work (78%), data processing (69%), data collection (64%), unpredictable physical work (25%), service work (20%), knowledge-based work (18%) and management (9%). Those workers engaged in routine work, who are unable to adjust to other types of work, will most likely become what Shipler terms "the working poor" (Shipler, 2005), or become part of the new "precariat" (Standing, 2014a, 2014b).

"Front-line workers" in the Figure 3.2 typology are all those workers in the service and healthcare sectors, including, for instance, workers in hotels, nursing homes, home helpers care, and so on. These workers will have jobs in the future, but with low income (Standing, 2014a, 2014b; Shipler, 2005; Gershuny & Fisher, 2014).

The typology also includes the category "thinking for a living" (Davenport, 2005), which will constitute knowledge workers with specialist skills (Hannah et al., 2015). These are people with highly specialized educations, at PhD level as a rule (Hanson, 2016). In all probability, this category will also consist of workers in several sub-categories, some of whom will be highly paid, while others will have middle-incomes (Ross, 2016). Most workers within this category will have work involving one or more of the twelve technological trends described by Kelly (2016) (see below). This category will involve work activities where robots and informats will play a large role. Despite the large degree of variation in the category, one common denominator will be that people will largely control their own use of time, while the workers in the other three categories of Figure 3.2 will have work that is time-governed (Lima, 2017, Hanson, 2016).

Innovation: herd behaviour and economic bubbles

One may ask: In what areas will future innovations emerge? The answer to the question can partly be found in Drucker's (1994: 44) analysis. Johannessen (2016: 153) has designed a model that provides an answer to the question. Future innovations will cause "disruptions" (Christensen, 2016), as well as small and large financial crises in the areas where productivity and quality are declining, where the real and relative costs are high, where the degree of diffusion of innovations spreads fastest, and in areas where new knowledge has the potential to become transformed into new technology. If this is correct, then it is highly likely that healthcare will be one of the major sectors in the future that will experience radical and possibly revolutionary innovations. As mentioned above, the large hospital complexes of today will probably be largely replaced by compact structures equipped with robots and informats that use the global infostructure to perform complex diagnoses and surgical procedures at low cost and of an extremely high quality.

The spirit of the times is innovation, cost control and productivity. When everyone is focused on innovation, there will be heavy investment in new opportunities by investors, and entrepreneurs will be attracted to this new area of activity. A form of "herd behaviour" will emerge in the market, figuratively speaking; i.e. resembling the behaviour of a flock of sheep or birds. This herd behaviour may easily become a problem, creating small and large bubbles in the economy. The rationale is that many will be simultaneously attracted to one or more innovations. When these economic bubbles burst, they will create minor and major economic crises, depending on the extent of the herd behaviour and how large the bubbles are. The hypothesis is that small innovations create minor economic crises, while major innovations create major economic crises (Johannessen, 2016).

One of the results of the herd behaviour will be overproduction within the area or areas that attract interest. Overproduction will lead to falling profits and bankruptcies. In this way, the system will clean itself. The next hypothesis is that the innovations that lead to the crisis will also be the same innovations that lead out of the crisis (Johannessen, 2016).

Another result of the herd behaviour is that radical innovations will lead to what Christensen (2010, 2016) terms disruption (as mentioned above). This leads to disintegration of the old business models and the emergence of new ones. Internet commerce is one such radical innovation that has destroyed old ways of doing business while simultaneously creating new businesses with completely different organizational structures. It is in this context that innovation leads to minor and major economic crises. If entire industries are transformed in a relatively short time, then people will lose their employment. Their expertise may become redundant. They lose their source of income and their futures do not look particularly bright. For some industries, this may take some time, but for others it can go quickly, very quickly. For instance, in the 1970s many of the textile workers in the OECD area were made redundant in less than one year (Hiemstra-Kupeus & Van Voss, 2010).

If it is the case as Davenport (2005: 1–11) writes that approximately half of those employed in the industrialized countries will lose their current jobs as a result of new technology, this will inevitably lead to minor and major economic and social crises. Of course, one can argue that in the longer perspective the technical evolution involving robots, nano-technology products, informats and other new technology will serve all (Abd, 2017; Wilson, 2017). In this context, one might also refer to Keynes' statement that "In the long run we are all dead". In the micro-historical perspective, the new technology will result in major upheavals for many, or, as Davenport writes, at least half of the workforce.

Since 2000, large parts of the traditional industry in the West has been moved offshore to other areas around the globe, such as China, India and other low-cost countries. In the initial developments, this concerned labour-intensive industries. Subsequently, a lot of high-tech manufacturing was also outsourced because costs were lower and profits higher. The new technology, especially that involving robots and informats, will result in many of the knowledge workers, those with higher education, losing their jobs (Ford, 2016; Beaudry et al., 2013). Many knowledge workers will become socially declassed and have to settle for jobs with lower salaries, greater uncertainty and poor expectations regarding future prospects (Bauman, 2013; Wacquant, 2009a, 2009b; Barrat, 2015).

The fact that the two expressions "the working poor" (Shipler, 2005) and "the precariat" (Standing, 2014a, 2014b) are gaining more attention perhaps reflects the situation that the outlook for many is very poor. Amongst the knowledge workers, the following occupational groups: journalists, lawyers, financial analysts, doctors, marketers, people in the advertising industry, etc. will experience that their jobs will be threatened by robots, informats, informatization and artificial intelligence (Brynjolfsson & McAfee, 2014; Gans, 2016). Moreover, all the routine-based and repetitive tasks, both in manufacturing and service-oriented professions, will also be threatened (Lima, 2017; Gershuny & Fisher, 2014).

While the industrial revolution increased agricultural productivity, the knowledge revolution increases the productivity of all types of knowledge workers (Drucker, 1999a, 1999b). The long view is that workers first migrated from agriculture to industry, and now from industry to knowledge occupations (Hannah et al., 2015; Hanson, 2016). While automation was industrialization's driving force, informatization is the driving force behind the knowledge-revolution. Informatization and informats are increasingly driven by cognitive robots based on artificial intelligence and interconnected through global technological networks (Wilson, 2017; Abd, 2017; Lima, 2017). As an extension of this, we see the emergence of cognitive technologies (Davenport, 2005: 4). Cognitive technologies may be understood as computers that observe, analyse, uncover patterns, create syntheses and make decisions (Ross, 2016; Wilson, 2017; Abd, 2017). For instance, your doctor, through instant access to cognitive robots and informats, may acquire knowledge that only the most skilled doctors in the world possessed previously (Davenport & Kirby, 2005: 4–5). This will apply in the near future also to car-driving, airline operations, financial analysis, legal evaluations, surgery, medical and psychological diagnoses, etc.

Cognitive technologies will enable professions to access the foremost global expertise in their fields through interconnected robots and informats that will create an open window of global knowledge. This knowledge window will obviously not be free, but will result in new competences and professions emerging. The resulting innovations will after a time-leg result in a situation where entrepreneurs and others flock to invest (Sassen, 2002; Sennett, 2013). This will result in herd behaviour encouraging the growth of an economic bubble, which when it bursts after a period will have consequences for individuals and social systems (Johannessen, 2016).

We have visualized this development in Figure 3.3.

The human brain will come into competition with artificial intelligence, because around 2026 robots will have the same processing capacity as humans (Lima, 2017; Abd, 2017). This means that all the work that is based on analysis and pattern recognition of large amounts of data will be able to be performed

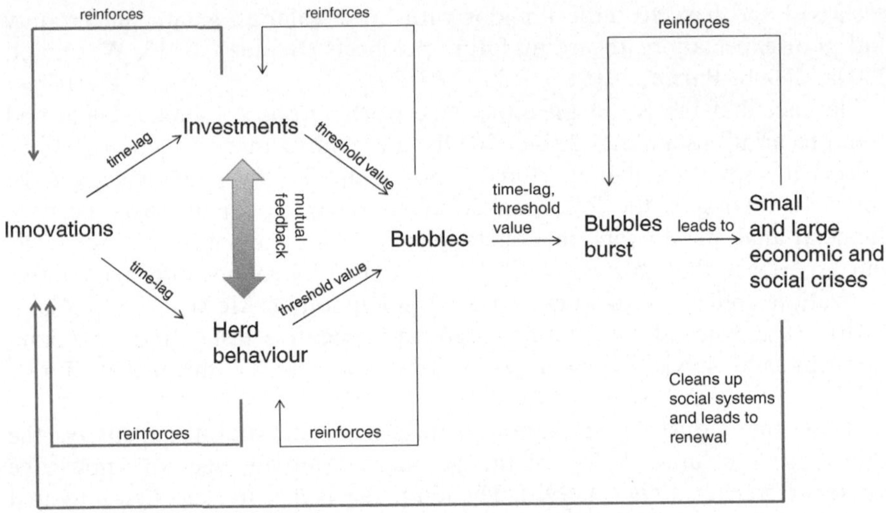

Figure 3.3 Herd behaviour, economic bubbles and economic and social crises

better by cognitive robots and informats than by humans (Wilson, 2017; Srinivasa, 2017).

In the short term, future innovations will disrupt the market leading to high levels of unemployment, and economic and social crises. However, as mentioned above, the same innovations that led to the crisis will, after a time-lag, also lead the way out of the crisis creating economic growth and rising prosperity for most people (Susskind & Susskind, 2015; Winfield, 2012). In the future, robots, informats and people will most probably work side by side; however, this will not be the case in the short term, because herd behaviour and economic bubbles will result in changes to the economy and social system that will lead to counter reactions against the new (Wood, 2016; Zuboff, 2017; Varoufakis, 2015). This also occurred during industrialization's childhood, when workers destroyed the machines because in many cases they replaced artisans and farm workers, making them redundant. At the beginning of the fourth industrial revolution, we can witness a similar reaction with the rising sentiment that "everything was so much better in the past". This may be understood as a nostalgia-reaction, which can have various economic, social and especially political consequences. For instance, Brexit, Trumpism and the emergence of national-conservative parties in Europe may be considered as elements of this "nostalgia" reaction.

Robots and informats will be the knowledge workers' future partners not competitors (Barrat, 2015). Just as the car was a mechanism that enabled geographic freedom, the robots and informats, and the new cognitive computers, will function as helpers enabling cognitive freedom (Bleuer et al., 2017). Robots and informats will initially outcompete some groups of knowledge workers (Bruce & Crook, 2015). In the mid-term and slightly longer term, the robots, robotization, informats and informatization will enhance our ability to increase productivity and quality, and result in new and more exciting work (Brynjolfsson & McAfee, 2014). Innovations may destroy established structures, but they also create long-term preconditions for economic growth.

Machines introduced into the English textile industry threatened the livelihoods of textile workers and weavers. Many of these workers banded together calling themselves Luddites;[3] with the aim of protecting their own jobs, they set about destroying the machines. The Luddites included workers in textile workshops, but also self-employed weavers in the cottage industry (Henty, 2016). The textile machines were an innovation in the industry, but also represented a "crisis" for those who lost their jobs, i.e. factory workers and cottage-workers. Their traditional skills were made redundant, and they lost their income. However, in the medium term, productivity of mills increased and new workers were hired into the industry, but with other functions than the traditional textile workers. Thus, in the short term, innovations led to personal and social crises; in the medium term, the innovations created more jobs than they destroyed; while in the long term, the innovations created economic growth and prosperity for most people (Jones, 2006).

Just as there was a struggle against the introduction of textile machines, we have also witnessed a struggle against automation (Jones, 2006: 20–35). Based on this historical knowledge, it should not surprise anyone if a similar struggle was initiated against robots, informats and informatization. Robotics and informatization will affect service jobs, knowledge workers' jobs and the administrative layers in organizations. In other words, these are what one may call "middle class" jobs. Therefore, it is highly probable that the middle class will be affected by these developments and become socially declassed in the fourth industrial revolution; robotization and informatization may even contribute to a dissolution of the middle class (Davidow & Malone, 2014). It is understandable that people are worried that robots and informats will take over their jobs. In this situation, it is the knowledge workers who have the most to fear (Brynjolfson & McAfee, 2014). Studies conducted in the United States show that the demand for knowledge workers reached its peak around the year 2000 (Beaudry et al., 2013). Concurrent with these developments, universities worldwide are spewing out yet more graduates qualified to work in the knowledge sector. Following the laws of economics, price is always a function of supply and demand, and consequently, knowledge workers' wages will fall in the future. The inequality between wages in various sectors will therefore increase (Piketty, 2014, 2016). Managers, financial analysts, investment funds analysts, investment bank employees, and so on, will be the salary winners (Spraque, 2015; Susskind & Susskind, 2015). The majority of knowledge workers will be salary losers (Davenport & Kirby, 2005: 6–7). Such developments indicate that extreme specialization will be the way to go to secure rewards. In the future, employees with regular university qualifications within the fields of law, medicine, engineering, economics and business administration will not receive particularly high salaries (Davidow & Malone, 2014). This means that robotization and informatization will create a new "working class" who will consist of knowledge workers with extensive education and relatively low salaries, and service workers with little education and low salaries. These developments will also be accompanied by the emergence of the so-called "precariat" (Standing, 2014a, 2014b) and "the working poor" (Shipler, 2005).

The "new working class" is characterized by a growing discontent. This dissatisfaction shows itself in many ways. In the US, Trumpism may be considered one "symptom" of these new developments. In Europe, Brexit and the emergence of national-conservative movements may be partly explained by various types of discontent. We are witnessing a rational dissatisfaction in relation to future prospects, an emotional dissatisfaction and political discontent. The political discontent affects the established politicians who are unable to provide answers to the problems and the challenges people are experiencing. The emotional dissatisfaction is related to the social declassification of some groups, and the fear of what the future may bring for people and their children. The rational discontent concerning future prospects may be linked to the idea that education should be worthwhile and lead to well-paying jobs. However, more and more people see that this is not the case. Master's degrees in fields such as psychology, medicine, law and other fields are perhaps not enough to provide meaningful, well-paid jobs; today, the job market often requires that graduates need to use a further five to ten years to specialize (Bruce & Crook, 2015; Gershuny & Fisher, 2014).

Destructive wealth creation: transformation of social systems

Innovation is the lifeblood of the modern economy, say Dyer et al. (2010: 1). The metaphor may be elaborated on by saying that some of this lifeblood leads to "bad blood" between people, because innovation leads to changes, and in some contexts transforms large parts of the economy, so that more people will experience social crises. Radical innovations transform whole wealth creation chains, and affect many people's lives (Christensen, 2010; 2016). After a time-lag, radical innovations create the new. Before the new is created, the old is destroyed. This is what is meant here by destructive wealth creation. While something is destructed, the new grows out of this destruction, like the phoenix bird rising from its own ashes. It is in such a context that Schumpeter (1951, 1954, 1989) uses the term creative destruction to describe what happens when innovations enter the market. Schumpeter makes the point that the creative upswing is related to the destruction caused by innovations. In other words, Schumpeter's concept creative destruction relates to a long-term perspective. Thus, innovations may be considered to have destructive and wealth creation elements. In the short term, the old is destroyed, while in the long term, the new is created. Destructive wealth creation relates to the short term, when economic bubbles develop and when they burst. Creative destruction relates to the upturn when the new is created and the old has been destroyed.

In the short term, innovations lead to bad blood between people, because they lead to new business models, while others are forced into bankruptcy; many people lose their jobs, some lose their income, and some people and families also experience social crises as a result of innovations. In the longer term, most people will benefit. However, this does not help matters in the short term for those who experience the worst consequences of small and large innovations. Those who are affected negatively by destructive wealth creation risk being degraded to a lower social class, while others are lying "wounded on the ground" and are in need of a blood transfusion from the lifeblood Dyer et al. (2011: 1) speak of. In such a situation, policies are needed to mitigate the most harmful effects of destructive wealth creation.

There are many historical examples of destructive wealth creation. The internal combustion engine replaced the steam engine; motor-driven ships replaced sailing ships, the refrigeration industry replaced other methods of keeping food fresh, such as salting; cars replaced horse-driven carriages, and so on. The social consequences of the above developments were in many cases considerable; some local communities were desolated, while others flourished and new ones joined and grew into large cities. Some businesses went bankrupt while others were started-up and prospered.

In recent times, we have examples of businesses that failed or ran into difficulties because they did not make the transition during changing times: for instance, Facit AB, the Swedish manufacturer of mechanical calculators who failed to make the transition to digital calculators; Kodak who did not see the new opportunities created by the infostructure and thus went bankrupt (Instagram took over many of Kodak's functional areas); Nokia who were unable to compete with the new smartphones that hit the market is another example (Skype now offers free telephone services); and Apple's iPad that outcompeted Sony Walkman, and so on.

Innovations create gaps in the market so the market is knocked out of balance. The greater the innovation, the larger will be the gaps, and, consequently, the greater will be the imbalance that occurs in the market. Entrepreneurs flock to exploit this gap in the market which creates opportunities for super-profits. The more entrepreneurs who try to fill the gap, the faster the rate of profit will fall, because competition forces prices down. After some time over-investment will occur due to this "herd behaviour". Economic bubbles are created, small and large. If there is no intervention at an overarching policy level the bubbles will burst, with the consequences this has for individuals, families, organizations and nations (Johannessen, 2016).

Destructive wealth creation changes the ground rules for entire industries, such as Amazon who changed the market of the book industry, and CreateSpace which made online publishing possible without editors as gatekeepers and intermediaries. Dell created an online model for direct sales of computing solutions; Rent the Runway offers for hire tailor-made designer pieces;

Research in Motion developed the "BlackBerry"; EBay developed online auctions between consumers and sellers; PayPal developed safe payment solutions over the Internet; Netflix, ViaPlay, etc. offer online movies; Uber established a global taxi business without owning a single taxi; Facebook have become the world's most popular news media without creating any of the news content; Airbnb do not own a single property, but have become a leading global provider of accommodation. All these companies have emerged as a result of the new technology and have created solutions that have led to large and small destructions of many established businesses and to economic and social crises; but they have also created wealth for new businesses.

One new aspect of the fourth industrial revolution will be that products will be transformed into services and processes (Kelly, 2016: 6–7). The "product" as we know it will be developed into a robot or informat that constantly updates information about our habits, desires and needs. Informats will be equipped with flexible technology that continuously adapts to user complexity. Consider a product such as a shoe. The shoe may be developed into a service product in the form of an informat that adapts to the user. For instance, consider a user with slightly different leg lengths after a hip replacement; the informat/shoe will be able to compensate for this. The shoe will also expand in the heat, become a winter shoe in the winter and a summer shoe in the summer, and so on. In other words, the shoe will become a service unit rather than a product (Kelly, 2016: 7). Products will be able to evolve into service units due to complex information processing (Case, 2016). These information processes will be connected to the individual service units, such as a shoe. The shoe will be connected to a global information system. If the user limps this will be rectified. If the user gets tired, the shoe will change shape. If the user develops an incipient disease, the shoe, and other service units the user uses, will have access to medical records, notify the appropriate entities, undertake measures and make corrections so the body is able to adapt and regain health balance. The ethical implications of such technology will be continuously discussed. If the individual does not want to be part of such a complex information system, he/she can disengage themselves from the information system.

Creativity is nothing new. However, what is new is that it will be the most sought after talent in the future (Dyer et al., 2010: 1). The reasoning is simple. Global competition entails competing on cost, quality and expertise, as well as innovation. Innovation provides the greatest profits, because the new can be priced higher than the monopoly price for a period. Innovation requires creativity. Therefore creativity will be the most sought after managerial skill in the future (Dyer et al., 2010: 261, note 1).

Change has become the new stability. Creativity is the new core competence. Just as you have finance departments in organizations that manage economic processes, in the future, creativity departments will most probably be established to manage creative processes. The reasoning is simple:

innovation is the area businesses will be able to stand out in the global economy, and creativity is a prerequisite of innovation.

The individual or business that is unable to adapt to changes will be made redundant. These changes will involve technology in one way or another. Technological changes transform established systems, creating new organizational structures and management roles (Brynjolfsson & McAfee, 2014; Brynjolfsson & Saunders, 2013). These technological changes, which will be connected to complex information systems, will be present everywhere. They will be in our computers, cars, clothes, shoes, pens, telephones, watches, jewellery, and so on. The information will be interconnected with global networks, where intelligence exists in interconnected structures of information networks (Catmull, 2014; Coates & Morrison, 2016). Which technological trends can we identify that will affect our future in the next thirty years? Kelly (2016) has a clear answer to this question, and proposes twelve technological trends that will be an integral part of our lives. We base the following trends on Kelly's description, but have created a different synthesis.

1 Technology of maintenance

Explanation: Everything changes over time (ref. The second law of thermodynamics).

2 Cognifying technology

Explanation: Smart freely available solutions will be developed based on artificial intelligence, robotics and informats that solve problems that we previously perceived as complex and intractable.

3 Technology of memory

Explanation: Everything that reaches the Internet is being copied, stored, retrieved, modified, re-copied and re-stored.

4 Screen technology

Explanation: Interaction via the written word in the book is replaced by interaction via various types of screens.

5 Loan and rent technology

Explanation: The trend is a shift from product to service, i.e. a technology that creates access without the need to own, such as Uber, Airbnb, Netflix, Photoshop, etc. Any product where customers have to wait, or costs are high, they develop their own "Uber for X", i.e. freelancers that can fill your needs.

6 Sharing technology

Explanation: The technology is free for all to use without charge, such as Skype, Wikipedia, etc. The trend is a wiki for everything, Creative Commons, Pinterest, Digg, Reddit, Tumblr, etc. This is a form of digital socialism where one can freely share what one has, so everyone can access it. Sharing technology is extremely decentralized. In the future, this technology will enable sharing opportunities in areas we are unaware of today, or we do not value.

7 Filtering and summarization technology

There is a constant barrage of new things, such as songs, books, blogs, online postings, and so on, so that it is impossible for both the man-in-the-street and specialists to keep up to date. Consequently, filtration and summarization technology is in demand to personalize and anticipate desires. Marketing based on people's searches on the Internet, for instance, Amazon, Google Books, Google Scholar, etc. You can also imagine a cognitive avatar that exists online, where your online searches are categorized and filtered to anticipate your desires; personalized products and services are then tested by the avatar before being sent to you.

8 Remixing technology

Explanation: When existing products are linked together and recombined entirely new solutions will arise that meet both old needs and new desires that have not yet been realized. Connecting established solutions will result in economic growth. For instance, consider, Snapchat, Flickr, WhatsApp, 3-D printers, and so on.

9 Synthetic technology

Explanation: This technology makes the experience more real than reality. For instance, holographic projections, avatars that are designed in solid material, a form of artificial world where dreams, needs and desires can be played out.

10 Personal diagnostic technology

Explanation: Watches and clothes that perform various health measurements, such as measuring your pulse, blood pressure, blood sugar level; or analysing your blood after consumption of food and beverages displayed on a screen so the individual understands how various types of food and liquids affect his/her body. This will give us better insight into how our health is affected by consumption and habits and give us the opportunity to change our behaviour. This technology may also be able to send signals directly to a robot or informat that decides courses of action. This technology will improve and be more accessible, and have a single focus: your health.

11 Mass editing technology

Explanation: Wikipedia was theoretically impossible, but practically tolerant of errors. Even the Supreme Court in the US have referred to Wikipedia articles. Everyone has the opportunity of developing and changing something in this type of wiki technology. One can imagine a situation where laws and regulations, diagnoses and medication may be developed in the same way: impossible in theory, but possible in practice. Everything that is now done by experts may be done in the future by a wiki tool that has been developed and edited by everyone and anyone, without exception.

12 Interactive technology

Explanation: Development of a "super-mind" on the planet Tellus that thinks and decides what is best for the whole and gives instructions to members (nations, institutions and businesses). This could conceivably be a materialization of the Gaia hypothesis, but now created by people. Such an interactive technology will be able to regulate the complex relationships that exist to preserve and maintain Tellus as a habitable planet where the whole and the parts are interconnected.

Our synthesis of the twelve technology trends are visualized in Figure 3.4.

Conclusion

The problem for discussion in this chapter was: How does robotization constitute an aspect of the fourth industrial revolution?

Robotization will lead to herd behaviour in the market and the growth of economic bubbles. After a given period, these minor and major economic bubbles will burst leading to minor and major economic, social and political

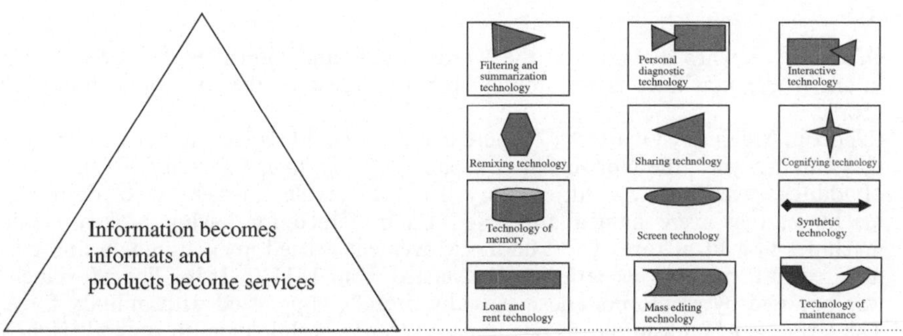

Figure 3.4 Technological trends that will create our future up until 2050

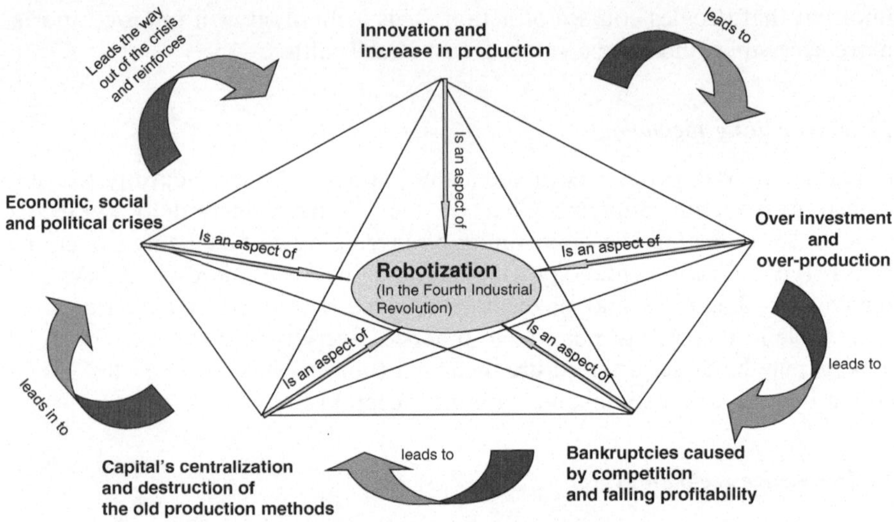

Figure 3.5 Robotization and the fourth industrial revolution

crises, and transformation of our social systems. However, the innovation that brings about a crisis will also be the innovation that shows us the way out of it. The time that elapses between the crisis and the upturn will be used to put things in order: to get rid of the old and bring in the new. This period is referred to as a crisis, but could just as well be packaged as a period of refurbishment to remove economic "tumours" and then start over with an economically healthy "body", cleansed of old dross. When using such a metaphor in an economic context, one should be aware that despite the long-term improvement of the economy's health, in a human context, many people, families and social systems will be adversely affected in the process.

We have developed an analytical model that visualizes the answer to the question posed in this chapter (Figure 3.5).

Notes

1 Of course, Keynes did not use the words robots and informats, but described a future where machines and automats perform most of the jobs done today by people.
2 Report in McKinsey Quarterly: "Where machines could replace humans–and where they can't … yet" (the Norwegian newspaper, *Aftenposten*, 3 January 2017).
3 "Luddites" were early 19th century English textile workers who destroyed machines; they were named after Ned Ludd, famed for having broken textile machines in a fit of rage. The Luddite Movement started in Nottingham and culminated in a region-wide rebellion that lasted from 1811 to 1816. The Movement was opposed by employers and eventually brutally suppressed with military force. The term Neo-Luddism means one opposed to new technologies in general (Henty, 2016).

References

Abd, K.K. (2017). *Intelligent scheduling of robotic flexible assembly cells*, Springer, London.

Barrat, J. (2015). *Our final invention*, St Martin's Griffin, London.

Bauman, Z. (2013). *Does the richness of the few benefit us all?* Polity, London.

Beaudry, P., Green, D.A. & Sand, B. (2013). *The great reversal in the demand for skill and cognitive tasks*, National Bureau of Economic Research, WP., no. 18901, New York.

Bleuer, H. & Bouri, M. (2017). *New trends in medical and service robots: assistive, surgical and educational robotics*, Springer, London.

Bleuer, H., Bouri, M. & Mandada, F.C. (2017). *New trends in medical and service robots*, Springer, London.

Brockbank, W. (2013). *Overview and logic*, in Ulrich, D., Brockbank, W., Younger, J. & Ulrich, M. (eds.), *Global HR competencies: mastering competitive value from the outside in*, McGraw Hill, New York, pp. 3–27.

Bruce, D. & Crook, G. (2015). *The dream café: lessons in the art of radical innovation*, John Wiley & Sons, New York.

Brynjolfsson, E. & McAfee, A. (2014). *The second machine age*, W.W. Norton & Company, New York.

Brynjolfsson, E. & Saunders, A. (2013). *Wired for innovation: how information technology is reshaping the economy*, The MIT Press, London.

Case, S. (2016). *The third wave*, Simon & Schuster, New York.

Catmull, E. (2014). *Creativity Inc.*, Bantam Books, New York.

Christensen, C.M. (2010). *disrupting class, expanded edition: how disruptive innovation will change the way the world learns*, McGraw-Hill, New York.

Christensen, C.M. (2016). *The Clayton M. Christensen reader*, Harvard Business School Press, Boston.

Coates, K.S. & Morrison, B. (2016). *Dream factories*, Dundum, London.

Davenport, T. H. (2005). *thinking for a living, how to get better performance and results from knowledge workers*, Harvard Business School Press, Boston.

Davenport, T.H. & Kirby, J. (2005). *Only human need apply: winners and losers in the age of the smart machines*, Harper Business, New York.

Davidow, W.H. & Malone, M.C. (2014). What happens to society when robots replace workers? *Harvard Business Review*, 10 December.

Drucker, P.F. (1994). *The age of discontinuity*, Transaction Publishers, New York.

Drucker, P.F. (1999a). Knowledge worker productivity: the biggest challenge, *California Management Review*, 41, 2: 79–94.

Drucker. P.F. (1999b). *Management challenges for the 21st century*, Harper Collins, New York.

Dyer, J., Gregersen, H. & Christensen, C.M. (2010). *The innovator's DNA: mastering the five skills of disruptive innovators*, Harvard Business Review Press, Boston.

Ford, M. (2016). *The rise of the robots: technology and the threat of mass unemployment*, One World, London.

Frankish, K. (2014). *The Cambridge handbook of artificial intelligence*, Cambridge University Press, Cambridge.

Gans, J. (2016). *The disruption dilemma*, The MIT Press, Boston.

Gershuny, J. & Fisher, K. (2014). *Post-industrious society: why work time will not disappear for our grandchildren*, Center for Time Use Research, Department of Sociology, University of Oxford, Oxford.

Hannah, E., Scott, J. & Trommer, S. (2015). *Expert knowledge in global trade*, Routledge, London.
Hanson, R. (2016). *The age of Em: work, love and life when robots rule the world*, Oxford University Press, Oxford.
Henty, G.A. (2016). *Through the fray: a tale of the Luddite Riot*, Create Space, New York.
Hiemstra-Kuperus, E. & Van Voss, L.H. (2010). *The Ashgate companion to the history of textile workers 1650–2000*, Routledge, London.
Johannessen, J-A. (2016). *Innovation leads to economic crises: explaining the bubble economy*, Palgrave, London.
Jones, S. (2006). *Against technology: from Luddites to Neo-Luddism*, CRC Press, New York.
Kelly, K. (2016). *The inevitable: understanding the 12 technological forces that will shape our future*, Viking, New York.
Keynes, J.M. (1963). *Essays in persuasion*, WW. Norton, New York.
Lima, P.U. (2017). *Autonomous mobile robotics: a system perspective*, CRS Press, New York.
Miller, J.G. (1978). *Living systems*, McGraw-Hill, New York.
Piketty, T. (2014). *Capital in the twenty-first century*, The Belknap Press of Harvard University Press, Boston.
Piketty, T. (2016). *Chronicles: On our troubled times*, Viking, London
Reich, R. (2015). *Saving capitalism: for the many, not the few*, Alfred A. Knopf, New York.
Ross, A. (2016). *The industries of the future*, Simon & Schuster, London.
Sassen, S. (2002). *Global networks/linked cities*, Routledge, New York.
Schumpeter, J. (1951) *Theory of economic development*, Harvard University Press, Boston.
Schumpeter, J. (1954). *History of economic analysis*, Oxford University Press, Oxford.
Schumpeter, J. (1989). *Business cycles*, Porcupine Press, New York.
Sennett, R. (2013). *Together*, Penguin, New York.
Shipler, D. (2005). *The working poor*, Vintage, New York.
Spraque,S. (2015). What can labor productivity tell us about the US economy, *US Bureau of Labor Statistics, Beyond the numbers* 3, 12 (May).
Srinivasa, R. (2017). *Whose global village: rethinking how technology shapes the world*, NYU Press, London.
Standing, G. (2014a). *A precariat charter*, Bloomsbury, London.
Standing, G. (2014b). *The precariat: the new dangerous class*, Bloomsbury Academic, New York.Susskind, R. & Susskind, D. (2015). *The future of professions: how technology will transform the work of human experts*, OUP, Oxford.
Vadakkepat, P. & Goswami P. (eds.). (2018). *Humanoid robotics: a reference*, Springer, London.
Varoufakis, Y. (2015). *The global minotaur*, Zed Books, London.
Wacquant, L. (2009a). *Punishing the poor*, Duke University Press, London.
Wacquant, L. (2009b). *Prisons of poverty*, University of Minnesota Press, New York.
Wilson, M. (2017). *Implementation of robot systems*, Butterworth-Heinemann, New York.
Winfield, A. (2012). *Robotics*, Oxford University Press, Oxford.
Wood, E.M. (2016). *Democracy against capitalism*, Verso, London.
Zuboff, S. (2017). *Master or slave: the fight for the soul of our information civilization*, Public Affairs, New York.

4 Aspects of a policy architecture for the fourth industrial revolution

Introduction

The fourth industrial revolution is the result of increased education, new technology, globalization of the economy, politics, culture and human relations (Schwab, 2016; Abd, 2017).

To investigate policy implications in the fourth industrial revolution, we have chosen an approach where we examine the following four subsystems (Bunge, 1998):

- The economic subsystem, focusing on material and technological resources.
- The political subsystem, focusing on power and distribution of resources.
- The collaborative subsystem, focusing on relationships and networks.
- The cultural subsystem, focusing on values and norms.

There is much to suggest that the rationality of the economic subsystem, i. e. the ideas and thinking concerning economic and technological issues, has impacted the cultural subsystem, the political subsystem and the relationships between people (Dorling, 2015; Chomsky, 2016a; Case, 2016).

The present management structure, as we know it, has been largely developed in relation to industrialization. However, when industrial workers become technicians who control robots and informats[1] a new type of management will emerge. This will also occur in traditional industrial companies (Brynjolfsson & McAfee, 2014).

The industrial society is associated with values and norms other than those prevailing at the start of the fourth industrial revolution (Castells, 2009; Chomsky, 2012). Central to the industrial society was the growth of trade unions and the focus on workers' rights. Additionally, the concept of fairness was a central value: this was understood to equate with the fair distribution of material resources, equal access to cultural institutions, equal participation in political life and strong resistance to the master–servant distinction in every sense (Dickinson, 2016; McGill, 2016).

At the beginning of the fourth industrial revolution, a major challenge is to incorporate in collective solutions respect and responsibility for, and the dignity of, the individual (Piketty, 2016; Roat, 2016). Another challenge will be to integrate productive capital and knowledge capital, while balancing financial capital so that it promotes sustainable value creation and does not operate within short time horizons which may damage knowledge-building in society as a whole (Rodrik, 2011; Rojecki, 2016).

While the prevailing normative basis in the industrial society was to ensure social and collective solutions, there is much to suggest that the normative basis at the start of the fourth industrial revolution will largely be concerned with ensuring individual solutions (Rosa & Mathys-Treio, 2015).

Globalization is not a result of democratic processes in the political subsystem. It may be better understood as an emergent dimension that has grown through interactions between the economic, political, cultural and collaborative subsystems. In the industrial epoch, it was largely the political subsystem that governed and controlled activities in the economic subsystem. However, at the beginning of the knowledge society and the fourth industrial revolution, characterized by globalization, the governing instruments of the national state's political subsystem do not have the same impact (Sennett, 2003).

At the organizational level, this development has led to a high focus on costs in order to remain globally competitive. In practice, this has led to global businesses and companies avoiding domestic taxation by using tax havens (Beck, 1992: 19; McGill, 2016). Consequently, if this development continues and increases the social contract between businesses, employees and nations may be worsened (Beck, 2015).

If the funding of collective resources is unevenly distributed in favour of businesses and capital income, this will probably increase social conflicts at the national level. This may reinforce the idea of "us against them" – thus encouraging a scapegoat mentality (Bauman, 2013). In its consequences, this may lead to the loosening of social ties in the nation state. Consequently, globalization may result in the loosening of social ties between businesses, workers and the national state. In addition, the idea of loyalty between businesses and employees may become less important. The labour market may also become more flexible of necessity, where workers will be expected to be available depending on the needs of employers. This "flexibility" will greatly reduce established working conditions. If globalization manifests itself in this way, this will probably result in new forms of collaboration and new collaborative networks that we can only guess the consequences of. If such a development takes anchor, there will be a loosening of social ties and a process of social disintegration will slowly develop. In turn, this will reinforce the trend of individualization that we can already see the contours of today in the competitive global economy (Sennett, 2003, 2013; Bauman, 2013).

The following developments may occur: global businesses will push down costs and increase profits. National budgets will be put under pressure because of a reduction in national income, while there will be an increasing demand for welfare services to be maintained or improved. Thus, although global businesses will make greater profits there will be increased pressure on social welfare systems, probably resulting in an even greater difference between the rich and poor (Mason, 2015; McGill, 2016).

In this introduction, we have attempted to describe the following research problem: the fourth industrial revolution may develop in such a way that it may be detrimental to organizations and nations.

To address this research problem, we have formulated the following main research question: Which overall policy can be developed so that the fourth industrial revolution will lead to greater value creation for the benefit of an increasing number of people?

In order to answer this research question we have formulated the following four sub-research questions:

Sub-research question 1: Which policy can be developed for the economic subsystem so that the fourth industrial revolution will lead to greater value creation for the benefit of an increasing number of people?

Sub-research question 2: Which policy can be developed for the political subsystem so that the fourth industrial revolution will lead to greater value creation for the benefit of an increasing number of people?

Sub-research question 3: Which policy can be developed for the cultural subsystem so that the fourth industrial revolution will lead to greater value creation for the benefit of an increasing number of people?

Sub-research question 4: Which policy can be developed for the collaborative system so that the fourth industrial revolution will lead to greater value creation for the benefit of an increasing number of people?

We have summarized the introduction in Figure 4.1, which also shows how the rest of the chapter is organized.

The economic subsystem

In this section we investigate the question: Which policy can be developed for the economic subsystem so that the fourth industrial revolution will lead to greater value creation for the benefit of an increasing number of people?

In the following, six elements will be discussed as aspects of a policy for the economic subsystem in the fourth industrial revolution. The six elements are: democratization of the economic subsystem, economic equalization, control over one's daily life, directing competition policy, the organized market and unique competence.

Democratization of the economic subsystem

Democratization of economic activity is important because this will increase engagement and motivation, thus promoting value creation (Charnock & Starosta, 2016). On the other hand, if financial capital is given free rein this

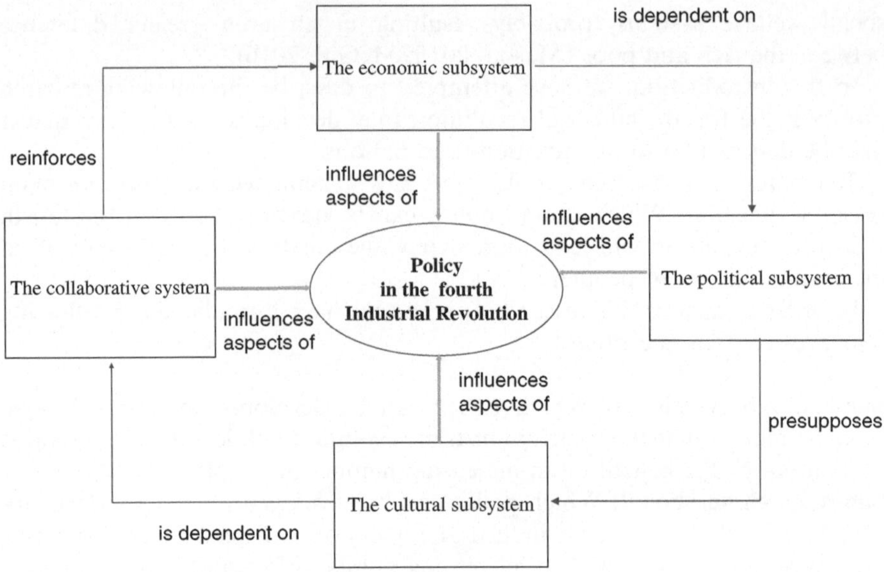

Figure 4.1 A policy architecture for the fourth industrial revolution

will result in a centralization of power and capital in the hands of the few (McGill, 2016; Bauman, 2013).

Directed competition policy and an organized market will be able to function as balancing mechanisms. Through these social mechanisms, engagement and motivation will promote value creation, while competition will function as a driving force for increased prosperity (Sennett, 2013; Rosa & Mathys-Treio, 2015). However, without these balancing mechanisms financial capital and uncontrolled competition could result in tearing the system apart (Bauman, 2013; Beck, 2015). Such a development would probably lead to a new type of capitalism emerging, namely feudal capitalism. Consequently, like Zappfe's paradox,[2] unrestrained competition will probably result in the self-destruction of the system.

Economic equalization and control over one's daily life

The purpose of economic equalization is to both ensure the welfare of the individual and to provide the basis for the social contract (Piketty, 2014: 1–33). If the social contract is eroded, the basis for social systems will also slowly disappear, paving the way for destructive individualism (Sennett, 2003; Roat, 2016).

Regarding economic equalization, output control is of interest because it focuses attention on the desired results (Hills et al., 2009). A focus on results should also be the position emphasized on all levels in the economic system, because it promotes responsibility, participation and greater control over one's daily life (Reich, 2015; Antonelli, 2001).

Setting standards for what is desirable at various levels promotes people's control over their own daily lives in relation to the economic processes they participate in (Charnock & Starosta, 2016).

Output control is particularly important in the knowledge economy because it stimulates participation and commitment (Dickinson, 2016). Output control as it is used here refers to having control of the results and what is delivered. However, output control must not be confused with goal management, which is something quite different. Output control is analogous with Wassily Leontief's Input-Output Model,[3] Stafford Beer's Theory of Living Systems (Beer, 1979, 1981, 1995) and Drucker's discussion of what motivates knowledge workers (Drucker, 1999a, 1999b). Knowledge workers differ from industrial workers regarding this point, i.e. having the possibility of controlling their own daily life (Gans, 2016; Gollan, 2010). This refers to control over your own use of time: not what you have to do, just where and when to do it. It is easier for the knowledge worker when the emphasis is on output control rather than process control (Bennis et al., 2012; Biswas-Diener, 2011; Boxall & Purcell, 2010).

It is always important to focus on the balance between stability and change, change and conservation, order and chaos, regulation and market, security and risk, because this balance helps to ensure economic equalization (Castelfranchi, 2007; Davies, 2003: 231–240). However, this requires an active government that focuses on the above, and responsible citizens who can help facilitate change and innovation processes (Harrison & Freeman, 2004; Lele, 2011).

An active focus on the three elements – the democratization of economic life, economic equalization and control over one's daily life – will counterbalance one of the greatest financial consequences of globalization: namely, the centralization of financial capital (Harvey, 2007; Mason, 2015). The centralization of financial capital is characterized by acquisitions, mergers, mergers of both finance centres and centres for managing real capital (Avent, 2016; Bolanski & Chiapello, 2017).

Economies of scale are viewed as having benefits such as cost saving and the concentration of power; however, the disadvantages are the social and economic consequences (Chomsky, 2016a, 2016b).

Directing competition and the organized market

If hyper-competition is the fundamental regulatory principle of economic life, this will probably lead to the emergence of a raw and brutal capitalism (Gupta et al., 2016; Harvey, 2007). Instead of such hyper-competition, which has been advocated by the neoliberalists since the 1980s, it is possible to imagine a policy of directed competition and an organized market (Petras et al., 2013; Avent, 2016). Directed competition balances the various negative factors in the so-called free market where hyper-competition prevails (Petras & Veltmeyr, 2011; Banki, 2015). Directed competition requires framework conditions that are relatively durable. Possibly, the most important framework conditions in directed competition are the collective agreements, laws and regulations that regulate the relationships and consequences of economic activities (Bolanski & Chiapello, 2017).

In directed competition the idea is that the market must be tamed, just as a wild horse needs to be tamed. When a horse has been tamed it can perform its tasks without harming anyone. However, if the horse remains untamed it may cause harm and be of no use to its owner.

Directed competition leads to an organized market. The illusion of the "free" market is propagated so that the few can benefit from this so-called freedom by enjoying the fruits of the labour of others (Chomsky, 2016a, 2016b).

On the other hand, a state regulated market is a historical anachronism and is viewed as contributing, amongst other things, to the fall of the Soviet Union and the collapse of Eastern Europe. However, directed competition and the organized market can use energy to "tame the wild horse".

For several thousand years, the market has been the starting point for trade and value creation. The organized market balances the various trends in the market, maintains competition and promotes value creation for the benefit of an increasing number of people (Pilger, 2016; De Sario, 2007).

Unless global market forces are tamed, the social consequences of free market thinking and hyper-competition may lead to self-destruction of the system as mentioned above in the reference to Zappfe's paradox, i.e. it's what you are good at that will be your downfall. Hyper-competition needs to be directed and the market organized, precisely so as to promote freedom, competition and market thinking. If balancing mechanisms are not installed to control some of the competition processes, then competition will cease to work (Dorling, 2015; Gant, 2014).

Global hyper-competition without global balancing social mechanisms may in its consequences lead to a sneaking introduction of various types of mandarin governance, and the development of a new type of capitalism, i.e. feudal capitalism. This will probably lead to an economic El-Nino literally causing the destruction of people's livelihoods and homes around the globe (Chomsky, 2016b; Ikonen, 2015). To use another analogy, giving free rein to financial capital may be likened to letting a fox into the henhouse.

Unique competence

One of the most important driving forces in the fourth industrial revolution is technological cutting edge expertise (Roat, 2016). It will therefore be extremely important for individuals, groups, organizations and nations to develop specific areas of expertise in which they possess unique competence. In the global economy, mediocrity, regardless of the level of education, will not constitute a competitive advantage. The skills that are developed must continually match what is demanded globally, because global demands will always materialise locally: i.e. the glocal (global + local) economic impact of the fourth industrial revolution.

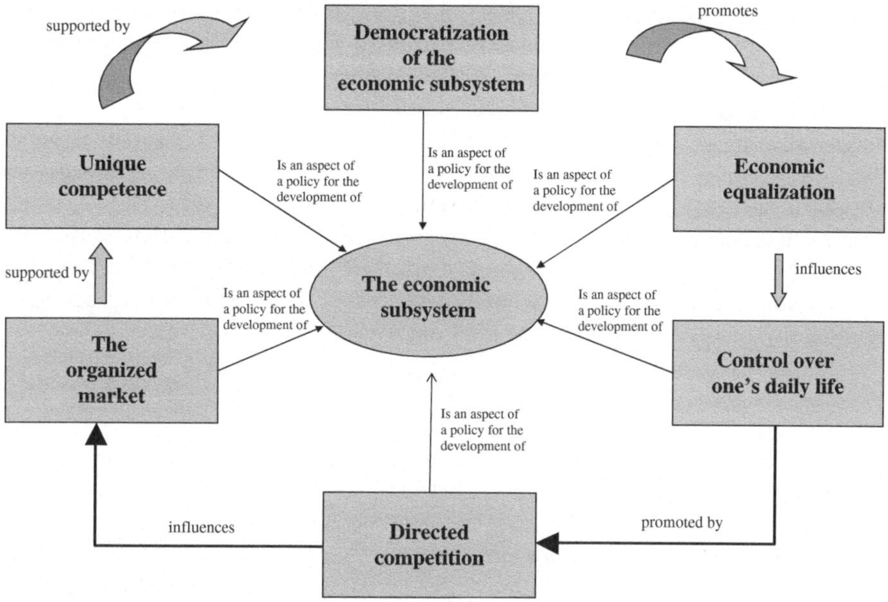

Figure 4.2 Aspects of a policy for the economic subsystem in the fourth industrial revolution

Summary

The economic subsystem's policy in the fourth industrial revolution is oriented around six elements described above and systematized in Figure 4.2.

The political subsystem

In this section, we investigate the question: Which policy can be developed for the political subsystem so that the fourth industrial revolution will lead to greater value creation for the benefit of an increasing number of people?

In the following, three elements will be discussed as aspects of a policy for the political subsystem of the fourth industrial revolution. The three elements are: power to the front-line, freedom from violation and activation requirements.

Power to the front-line and freedom from violation

It is incorrect to argue that it is the democratic system that is under pressure in the global knowledge economy (Johnson, 2015; Kessler, 2017). It is rather people's lack of influence over the processes that affect their everyday lives leading to their passivity and lack of engagement (Lazzarato, 2012; Mason, 2012, 2015). Such an understanding leads to completely different action strategies and policy design than the assumption that democracy is under any kind of pressure.

New developments that increase the individual's participation in the processes that affect him/her will also result in an increase in their participation in democratic processes (Tarrow, 2005; Tolonen, 2005). It is being able to influence the processes affecting one's daily life that are of interest here, not necessarily the fact that the processes are local (Trot, 2015). Participation in the local arena is not necessarily more democratic than being able to exert an impact in distant decision-making forums that affect one's daily life. In other words, democracy does not necessarily concern something that is locally located but rather is to do with the potential of influencing the processes that affect one's life, regardless of what the processes may be or if they are geographically located at a distance. This requires at least two things:

1 The technological opportunity to participate in decision-making processes that affect one's daily life.
2 Trust in the fact that people know what is best for them, and then acting in accordance with what they believe.

Thus, it is not a question of near or distant geographic location but greater opportunity to influence one's daily life, no matter where the processes have their geographical location (Varga, 2015; Savage, 2015). If one manages to develop such democratic processes, then there is the possibility that this will reduce the instances where people feel violated (Bennis et al., 2012).

One consequence of globalization is "tribalism", i.e. emotional commitment to that which is local, ethnic identity, patriotism, nationalism, the individual's own roots, myths, local narratives and knowledge of one's own history, understanding of one's language and culture, and the personal (Reich, 2015; Wiedemer et al., 2015). It is in this context that freedom from perceived violation must be given priority before the said violation takes undesired directions (Avent, 2016; Johnson, 2015).

Tribalism is based on solidarity between certain groups of people, where the loyalty is to the group and not to the nation or global structures (Rozenblit, 2008). Democratic developments in various countries may be influenced by the activities of these groups. Examples of this are the French-speaking community in Canada; the Northern Ireland Conflict; the Basques in Spain; the Kurds in Turkey, Iran and Iraq; the Spanish-speaking community in the US; the Christians in Southern Sudan, etc. The identity that tribalism promotes can shatter existing structures at the national level and affect the development of collaborative structures at the global level (James, 2006). New technology, including robotization, digitalization, informatization and automation, may reinforce the focus on the local, the group and tribalism, i.e. on "us against them" (de Anca, 2012). This may be explained as viewed from two perspectives: firstly, because cultures are being globalized, thus promoting local involvement (Nordberg, 2016), and secondly, because the new technology enables strong structural connections (Brynjolfsson & McAfee, 2014). These structural links can constitute a social mechanism for the rapid mobilization of local involvement. There are at least two types of mobilizations: that which rises to the surface and is visible, and that which does not show itself but remains more hidden. The hidden mobilization enables the maintenance of the feeling of "belonging to the tribe", because contacts can be maintained in new ways. The new technology enables structural links and identity across geographic distances because identity and background are similar (Nordberg, 2016). With such a backdrop, it is reasonable to assume that respect for the individual and the safeguarding of the dignity of the individual, as well as responsibility for others, will be strengthened. Such a development will, in all likelihood, strengthen people's perceived dignity, reduce perceived violations and mitigate undesired actions (Benhabib, 2002, 2004; Benhabib et al., 2006).

The activation requirement

It is not a desirable solution to pay people to be passive, thus excluding them from social responsibility, simply because they for some reason are unable to fulfil what is a negotiated norm, such as an 8-hour work day (Bolanski & Chiapello, 2017).

The more active people are in relation to the processes that influence their own daily lives, the greater the likelihood of real power over their own lives (Beck, 2015). Activating those who are affected by the consequences of political processes will become a principal political task in the global knowledge economy (Sennett, 2003, 2013). However, activation will not necessarily be realized as a consequence of increased decentralization (Beck, 2015). In this context, the question is whether decentralization is of any real significance or whether there are other processes that have a greater impact on people's lives. It is the activation of those who are affected, which is of interest in the democratization of decision-making processes in the fourth industrial revolution (Bennis, et al., 2012). If activation is not increased and people do not gain more influence over their own lives, centralization of power may fall into the hands of a small minority (Bauman, 2013; Beck, 2015). This will only increase the distance between people and those that govern them, between the political and economic elite and those who do not belong to the elite (McGill, 2016). It seems reasonable to assume that increasing democratization in the fourth industrial revolution will place greater demands on the individual's activation in exchange for social benefits (Sennett, 2003; Savage, 2015). Thus, in the near future it seems probably that the individual citizen will be required to participate actively in return for their social benefits (Beck, 2015; Sennett, 2013). This could reduce the gap between those who govern and those who are governed, and it will also contribute to lessening the conflict between the elite and the others.

Summary

The political subsystem's policy in the fourth industrial revolution is oriented around the elements described above and systematized in Figure 4.3.

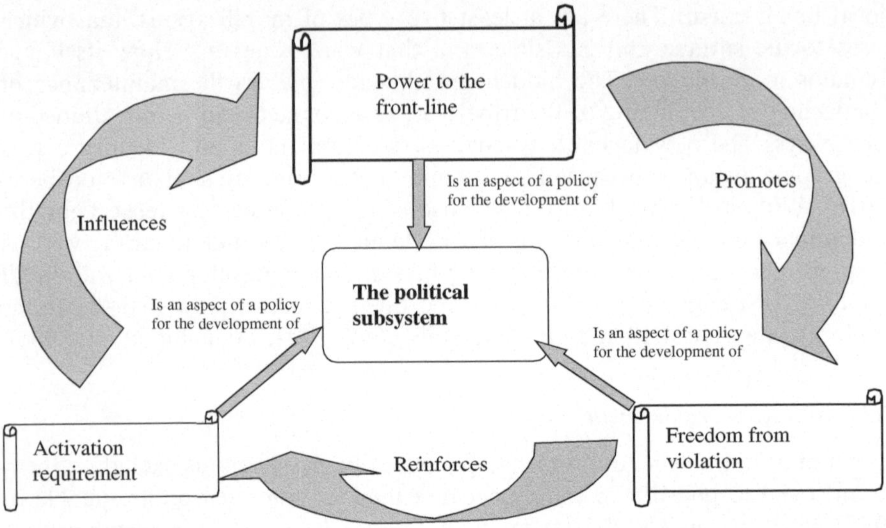

Figure 4.3 Aspects of a policy for the political subsystem in the fourth industrial revolution

The cultural subsystem

In this section we investigate the question: Which policy can be developed for the cultural subsystem so that the fourth industrial revolution will lead to greater value creation for the benefit of an increasing number of people?

In the following, three elements are discussed which are intended as aspects of a policy for the cultural subsystem in the fourth industrial revolution. The three elements are: respect for the individual, the dignity of the individual and responsibility for others.

Respect for the individual

In the beginning of the fourth industrial revolution, there is much to suggest that greater focus will be placed on respect for the individual (Benhabib, 2002, 2004; Benhabib et al., 2006).

In the industrial society, we often placed people in a waiting position. Figuratively this may be imagined to be in a warehouse where they were stored away like some machine part until they were needed; sometimes they would be taken out of storage, at other times people would be kept in storage their whole lives. If respect for the individual is to be a fundamental principle of the fourth industrial revolution, where competence is the most important social value creation mechanism, the above type of thinking that characterized the industrial society is both unethical (Benhabib, 2004) and of little use (Beck, 2015).

In the industrial society it was possible, perhaps even desirable, and in any case a common practice, to abandon what was considered to be "worn out" – parts that could be replaced with newer parts (Ackoff, 1989). Similarly, as mentioned, people were also treated like interchangeable machine parts (Bauman, 2011). However, if this way of thinking is adopted in the knowledge society, it may be likened to closing down half a library so that the books will not be worn out from being used. In this context, it is one's thinking that is problematic and not the actual phenomenon (Armano & Murgia, 2015: 102–117; Boxall & Purcell, 2010). The phenomenon in this case is people's competence. If this is to be transformed into value creation for the individual and society, it is a question of what is the most effective organization (Kessler, 2017).

When people are treated with respect and dignity, this forms the basis for social activation, involvement, participation and commitment (Benhabib, 2002, 2004; Sennett, 2013). Being active in cultural, political and economic value creation processes is a question of degree. This will depend on the individual's capabilities and background. It is this aspect that is of interest in the organization of the knowledge society (McAfee & Brynjolfsson, 2017). No one needs to be "stored away" for shorter or longer periods. It is rather a question of what is the most effective organization that can take full advantage of an individual's competence, although this competence may not fulfil certain "norms" such as the 8-hour work-day (Sennett, 2013).

In the industrial society, employees were active on the basis of a specific productivity norm. In the knowledge society, however, people can be active on the basis of a competence norm. Organized in this way, everybody will have the opportunity of being active on the basis of their background and capabilities, but never inactive. This will also result in the individual being treated with respect (Benhabib, 2004).

New developments suggest that identity appears to be veering away from the collective (Varga, 2015; Trot, 2015). The focus now seems to be on developing the skills needed for a particular job or task, but also the need of skills and knowledge to work with the new technology (Davenport & Kirby, 2005). In such a situation, it seems reasonable to assume that the social contract based on responsibility for the collective will become less prominent (Meister & Mulcahy, 2017: 1–15). More attention will be focused on the individual, and respect for the individual will be strengthened in such a context (Bauman, 2011). On the other hand, responsibility for, participation in and commitment to the collective will be given less emphasis (Castells, 2015). Although the focus will be more on the individual, it is not necessarily the case that the dignity of the individual will receive more attention (Charnock & Starosta, 2016).

If respect for the individual is to be given more emphasis, an action strategy may be to give the individual greater opportunity to choose his/her own future, although this may lead to collective solutions being put under pressure (Sennett, 2013; Beck, 2015).

The dignity of the individual

When unemployment becomes an integral part of the organization of society, responsibility for others is disregarded, and the dignity of the individual is violated (Chomsky, 2016a, 2016b).

It seems reasonable to assume that the perspective on unemployment is shifting in the knowledge society because competence has become the main input factor and the most important output factor (Morris, 2012; Nordberg, 2016). The same should also apply to the perspective on senior workers because their experience-based skills can be of importance as part of skills development in the institutions that develop skills, and in the systems that use this competence. Therefore, several ways of organizing experience transfer and knowledge building may be used where senior workers become part of this process. In this way, rather than being shuffled into retirement, senior workers will be able to actively participate in working life, albeit on different conditions than before. It is not only more rational to use the expertise of people that is available, no matter at which stage the person is in their life cycle, it will also show respect and strengthen his or her dignity by including them in key social activities (Sennett, 2013).

It lessens human dignity when people are hidden away in unemployment queues, receiving benefits on various social security schemes of imagined usefulness, and when healthy people are shuffled into retirement. Age is not a sickness, so those who are able and willing to work should be given the opportunity to participate in value creation activities in one way or another.

Responsibility for others

Flexibility is a significant dimension in the knowledge society (Pearson, 2015; Perlin, 2011). The reason is that competence is the key to value creation processes. Competence development is under continuous change in the fourth industrial revolution, and may therefore appear threatening to the individual (Rojecki, 2016; Avent, 2016). Therefore, individuals, organizations and society need to organize regarding values and standards in such a way that flexibility is a key factor for activities (Lele, 2011).

This flexibility can, amongst other things, be expressed in the way work is organized (Reinmoell & Reinmoeller, 2015). This could apply to people who suffer from a disability, who have reduced working capacity, and also in relation to senior workers (60+). It may be imagined that changing roles in working life could include senior workers, such as where they take on roles involving the coaching and supervising of younger workers. A mentor role would naturally replace a more active role for senior workers in the knowledge society.

Historically, unemployment has been a recurring phenomenon throughout the industrial era. This need not be the case in the knowledge society. Important in this context is flexibility during the course of one's working lifecycle regarding the development of knowledge, the application of knowledge and, towards the end of one's working lifecycle, the transfer of knowledge and experience.

People are not considered to be unemployed when at some point in their lives they are acquiring knowledge, such as in school or university – similarly, in the case mentioned above where senior workers (65+) transfer their knowledge and experience in roles as mentors or coaches, or perform other tasks. In the knowledge society, it is precisely competence that will be subject to continuous renewal because turbulence and complexity will increase. Such thinking is also linked to taking responsibility for others, as well as showing respect to seniors so that they can live worthwhile lives.

Summary

The cultural subsystem's policy in the fourth industrial revolution is oriented around the elements described above and systematized in Figure 4.4.

The social subsystem

In this section we investigate the question: Which policy can be developed for the collaborative system so that the fourth industrial revolution will lead to greater value creation for the benefit of an increasing number of people?

We discuss five elements in the following. These are: global competence clusters; collaborative networks; social inclusion; social justice; and social participation.

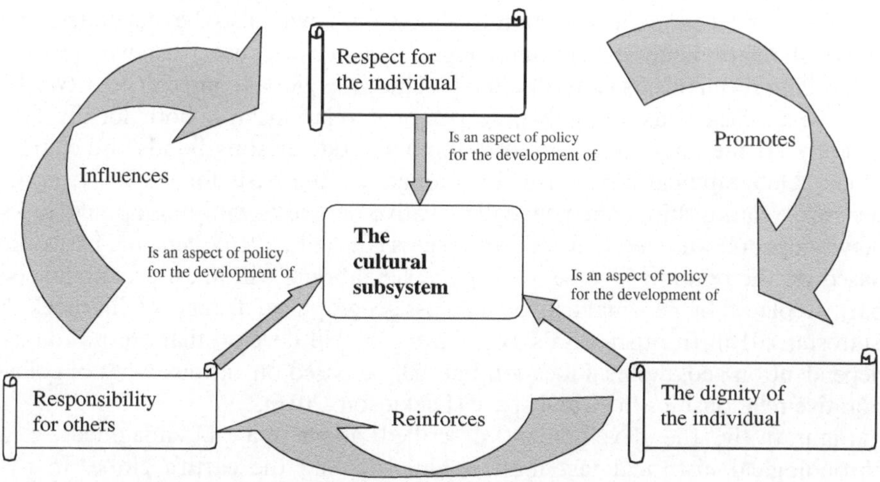

Figure 4.4 Aspects of a policy for the cultural subsystem in the fourth industrial revolution

Global competence clusters and collaborative networks

Firstly, a new aspect of global competence clusters is that they are distributed differently than previous international networks of various types (Abrahamson, 2004). Secondly, they are more structurally linked so that changes in one place more easily impact on other places in the network and/or in other global competence clusters (Avent, 2016). It is the new technology that facilitates and largely drives the global competence clusters (Wilson, 2017; Zhao et al., 2017). The global competence clusters may be likened to fishing nets lying on top of each other in a stack. Each fishing net constitutes a global competence cluster. The fishing nets are also in contact with each other along the vertical plane. This figurative image aims to elucidate how the separate global competence clusters are also structurally linked to other global competence clusters. In this way, it is possible to talk about the degree of internal structural links in the individual global competence clusters, and the degree of external structural links between the individual global competence clusters. The latter is illustrated by the connections between the fishing nets in the image described above.

Indeed, the better the external structural links are in the global competence clusters, the greater the level of expertise the network will have (Wilson, 2017). Global competence clusters will change the way we think, how we think, what we think, who we work with and the actions that result from this thinking and collaboration (Wakeling & Savage, 2015a, 2015b). The global competence clusters will form an organizational innovation in the global space, which will have an impact on value creation in the fourth industrial revolution (Susskind & Susskind, 2015).

Four types of collaborative networks may be envisaged (Bunge, 1998): the economic, political, social and cultural. The cultural collaborative networks will largely be based on personal relationships involving strong and weak ties. Similarly, the economic collaborative networks will also be dependent on personal relationships to strengthen performance because trust will replace controlling mechanisms (Castells, 2009). Thus, the global competence networks may be envisaged as strengthening personal relationships, both locally and globally. At the same time, globalization will erode existing bonds and change others. Geographical location will no longer be the basis for the fostering of personal relationships. The new collaborative networks will develop new links between people and new cultural expressions (Castells, 2015). Personal networks based on the relationships between good neighbours will in all probability be partly replaced by personal networks across geographical distances (Charnock & Starosta, 2016). In other words, social systems will develop that are no longer dependent on geographical location but will be based on various types of collaborative relationships in virtual space (Dickinson, 2016).

Figuratively, the development described above may be imagined as a chronological historical fast motion film showing the earth's globe. In the present and past, where the national borders exist, relationships and collaboration are etched in as dynamic dimensions changing gradually over time.

In the present and future, the fast motion film will show how national boundaries are softened with the passage of time and gradually replaced by networks of various types across borders. In terms of social consequences, new social systems will evolve over time based on these global clusters of competence. These may be envisioned as spreading out like waterfalls on this fast motion film across the aforesaid borders. If the development of various information and communication flows are also then etched onto this globe, a new reality would appear where the national borders are only of historical interest (Sennett, 2003).

Social relations have acquired a new character based on the new collaborative networks (Sennett, 2013). Therefore, new social systems will have other social matrices than those of the national state. How identity and loyalty are shaped may also then take on a different character (Castells, 2015).

Social inclusion, social justice and social participation

Solidarity in the knowledge society may be transformed into a more rational solidarity, i.e., solidarity will be viewed as something that directly benefits the individual (Sennett, 2013, Florida, 2014). This type of solidarity concurs with the increased individualization which we can observe at the start of the fourth industrial revolution (Goodman, 2015). However, individualization must not be confused with egoism. It seems reasonable to assume that individual solutions may also result in improved social justice. In the industrial society, the core of solidarity was also self-interest. The difference is that while rational solidarity is based on individual solutions, traditional solidarity takes as its starting point collective solutions. It is not, therefore, a weakening of solidarity that will emerge in the knowledge society – it will only become more explicit who it benefits (Bauman, 2011, 2013).

Clear solutions to complex issues appear to be the preferred approach, rather than complex solutions. This is because clear solutions will provide greater transparency, which will also facilitate social participation (Sennett, 2013; Castells, 2015). On the other hand, clear solutions should not be confused with simple and easy ones. Improved communication and system integration may be used to reduce complexity, reduce uncertainty about information, and minimize ambiguity (Luhmann, 1996; Bunge, 2015). Complexity may also be reduced by integrating all the systems operating within the specific problem area, such as the school, the family, the police and healthcare.

Intervention will involve systemic integration, i.e. responsibility for the part and whole of the person it concerns (Bunge, 1998). This may be done through integrating the entire life of the person in focus. It is precisely integration that is desirable to achieve the necessary intervention. The integration of the insular professional institutions and the structural links between the various professions will provide the opportunity to increase the quality of life of the individual. In this way, social participation will be encouraged because solution strategies will become transparent, connected and clear.

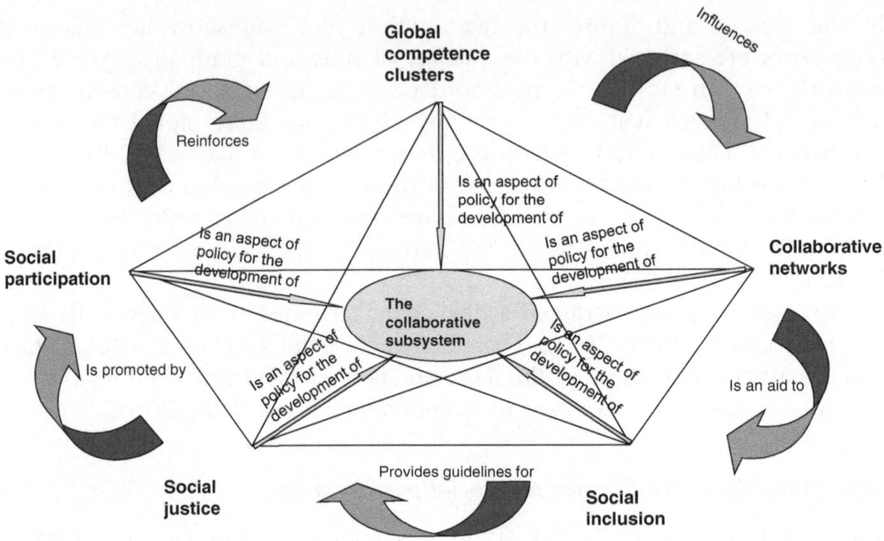

Figure 4.5 Aspects of a policy for the collaborative system in the fourth industrial revolution

Summary

Aspects of a policy for the collaborative system in the fourth industrial revolution will be oriented around the elements described above and systematized in Figure 4.5.

Conclusion

The research question we have investigated in this chapter is the following: Which overall policy can be developed so that the fourth industrial revolution will lead to greater value creation for the benefit of an increasing number of people?

In brief, we have designed 17 policy interventions, which comprises: six for the economic subsystem, three for the political subsystem, three for the cultural subsystem and five for the collaborative system. Based on the investigation in this chapter, we have illustrated the response to the research question in Table 4.1.

Theoretical and practical implications

Changes in bureaucratic management and control structures are considered to be a major challenge at the start of the fourth industrial revolution. Traditional management ideology is associated with the industrial society, and is

Table 4.1 Policy interventions in the fourth industrial revolution

Social subsystem	*Expressed through*	*Aspects of a policy for the fourth industrial revolution*
The economic subsystem	Material resources	Democratization of the economic subsystem Economic equalization Control over one's daily life Directing competition The organized market Unique competence
The political subsystem	Power	Power to the front-line Freedom from violation Activation requirement
The cultural subsystem	Values	Respect for the individualThe dignity of the individual Responsibility for others
The collaborative subsystem	Human relations	Global competence network Collaborative networks Social inclusion Social justice Social participation

neither possible nor desirable in the globalized knowledge economy. It is not possible because information and knowledge processes are largely driven by the front-line of the system. It is not desirable because it will suffocate initiative and creativity. However, this does not mean that management and control are not possible or desirable. It is possible precisely because self-organization enables management by those who have the expertise and operate in the front-line. It is desirable because the possible outcomes are infinite and complexity needs to be reduced.

It seems reasonable to assume that the economic aspect of globalization will lead to increased tensions, due to differences between profits and wages. This may occur in both the countries where the new industrialization is growing, and in the old industrialized countries that are being transformed into knowledge economies. The knowledge economies will increasingly base their value creation on robots, informats and artificial intelligence.

In the same way that we learned to live with uncertainty in the industrial society, we must learn to live with ambiguity in the knowledge society. Uncertainty was dealt with through using information. Ambiguity may be reduced by communication. Multicultural involvement will be an aid for learning how to deal with ambiguity, because the knowledge society operates in the global space, which by definition is multicultural. In such a global world, cultural identity becomes important. However, cultural identity must not be confused with cultural protectionism (Chomsky, 2016a, 2016b). Cultural protectionism can take many isolating and excluding forms. Cultural protectionism may easily descend into national protectionism that excludes those who do not "belong" in the national space. The worst forms of cultural protectionism belong to the darkest sides of European history.

Identity grows literally over time as a distinction between a system and its external world. It is important to be aware of which cultural identity one wishes to create and to show to others. Cultural identity is what distinguishes us from others. However, one should be aware that there is a delicate balance between distinguishing oneself from others and separating oneself from them. The latter is the expression of cultural protectionism.

Creolization, also known as cultural pluralism, is a balancing mechanism that can ensure that cultural identity does not descend into cultural protectionism. Creolization is the multicultural expression of respect, responsibility and dignity. Creolization offers protection against intolerance, oppression, violation and exclusionary attitudes.

In the knowledge society, competence is possibly the most important input factor for value creation. In order to attract competent people, you need a culture that is appealing. Consequently, culture may become the most important factor when attracting competence because it relates to the identity of others. In order to compete in the global market, it is crucial that social systems have the best skills within their defined area. Taking this into consideration, culture is not so much the result of the economic development of a country, but something that needs to be fostered so it appears as appealing and provides a basis for value creation.

Notes

1 Informats are robots which are globally interconnected. In this way they continuously acquire new information within the functional area.
2 The philosopher Zappfe formulated (in brief) the following paradox: It's what you are good at that will be your downfall
3 https://en.wikipedia.org/wiki/Input–output_model (access date: 28.3.2017). Leontief was awarded the Nobel Prize in Economics in 1973.

References

Abd, K.K. (2017). *Intelligent scheduling of robotic flexible assembly cells*, Springer, London.

Abrahamson, M. (2004). *Global cities*, Oxford University Press, Oxford.

Ackoff, R.L. (1989). *Re-creating the corporate future*, Oxford University Press, Oxford.

Antonelli, V. (2001). *The microeconomics of technological change*, Oxford University Press, Oxford.

Armano, E. & Murgia, A. (2015). The precariousness of young knowledge workers: a subject-oriented approach, in Johnson, M. (Ed.). *Precariat: labour, work and politics*, Routledge, London, pp. 102–117.

Avent, R. (2016). *The wealth of humans: work and its absence in the twenty-first century*, Allen Lane, New York.

Banki, S. (2015). Precarity of place: a complement to the growing precariat literature, in Johnson, M. (Ed.). *Precariat: labour, work and politics*, Routledge, London, pp. 66–79.
Bauman, Z. (2011). *Culture in a liquid modern world*, Polity Press, London.
Bauman, Z. (2013). *Does the richness of the few benefit us all?* Polity, London.
Beck, U. (1992). *Risk society: towards a new modernity*, Sage, London.
Beck, U. (2015). *Reinvention of politics: rethinking modernity in the global social order*, Polity, New York.
Beer, S. (1979). *The heart of enterprise*, John Wiley & Sons, Chichester.
Beer, S. (1981). *Brain of the firm*, John Wiley & Sons, New York.
Beer, S. (1995). *Diagnosing the system for organizations*, John Wiley & Sons, London.
Benhabib, S. (2002). *The claims of culture*, Princeton University Press, Princeton.
Benhabib, S. (2004). *The rights of others*, Cambridge University Press, Cambridge.
Benhabib, S., Waldron, J., Honig, B. & Kymlicka, W. (eds.) (2006). *Another cosmopolitanism*, Oxford University Press, Oxford.
Bennis, W.G., Cloke, K. & Goldsmith, J. (2012). *The end of management and the rise of organizational democracy*, John Wiley & Sons, New York.
Biswas-Diener, R. (2011). *Positive psychology and social change*, Springer, London.
Bolanski, L. & Chiapello, E. (2017). *The new spirit of capitalism*, Verso, London.
Boxall, P.F. & Purcell, J. (2010). An HRM perspective on employee participation, in Wilkinson, A., Golan, P.J., Marchington, M. & Lewins, D. (eds.). *The Oxford handbook of participation in organizations*, Oxford University Press, Oxford, pp. 129–151.
Brynjolfsson, E. & McAfee, A. (2014). *The second machine age*, W.W. Norton & Company, New York.
Bunge, M. (1998). *Philosophy of science: from problem to theory*, Volume one, Transaction Publishers, New Jersey.
Bunge, M. (2015). *Political philosophy*, Transaction Publisher, New York.
Case, S. (2016). *The third wave*, Simon & Schuster, New York.
Castelfranchi, C. (2007). Six critical remarks on science and the construction of the knowledge society, *Journal of Science Communication*, 6(4), 1–3.
Castells, M. (2009). *The power of identity*, Wiley-Blackwell, Oxford.
Castells, M. (2015). *Networks of outrage and hope*, Polity Press, New York.
Charnock, G. & Starosta, G. (2016). *The new international division of labour, global transformation and uneven development*, Palgrave, London.
Chomsky, N. (2012). *How the world works*, Hamish Hamilton, London.
Chomsky, N. (2016a). *Profit over people: war against people*, Piper, Berlin.
Chomsky, N. (2016b). *Who rules the world*, Hamish Hamilton, London.
Davenport, T. H. & Kirby, J. (2005). *Only humans need apply: winners & losers in the age of smart machines*, Harper Business, New York.
Davies, A.C. (2003). Are firms moving Downstream into High-Value Services? In Tidd, J. & Hull, F.M. *Service innovation: organizational responses to technological opportunities & market imperatives.* Imperial College Press, London, pp 321–340.
De Anca, C. (2012). *Beyond tribalism*, Palgrave, London.
De Sario, B. (2007). Precari Su Marte: an experiment in activism against precarity, *Feminist Review*, 87, 1: 21–39.
Florida, R. (2014). *The rise of the creative class*, Basic Books, New York.
Dickinson, E. (2016). *Globalization and migration*, Rowman & Littlefield, London.
Dorling, D. (2015). *Inequality and the 1%*, Verso, London.

Drucker, P.F. (1999a). Knowledge worker productivity: the biggest challenge, *California Management Review*, 41, 2: 79–94.

Drucker, P.F. (1999b). *Management challenges for the 21st century.* Harper Collins, New York.

Gans, J. (2016). *The disruption dilemma*, The MIT Press, Boston.

Gant, A. (2014). *Give and take, Why helping others drives our success*, W&N, New York.

Gollan, P.J. (2010). Employer strategies towards non-union collective voice, in Wilkinson, A., Golan, P.J., Marchington, M. & Lewins, D. (eds.). *The Oxford handbook of participation in organizations*, Oxford University Press, Oxford, pp. 212–236.

Goodman, J. (2015). *Crisis, movement, management: globalising dynamics*, Routledge, London.

Gupta, S., Habjan, J. & Tutek, H. (2016). *Academic labour unemployment and global higher education: neoliberal politics of funding and management*, Palgrave, London.

Harrison, G. & Freeman, R. (2004). Democracy in and around organizations, *The Academy of Management Journal*, 18, 3: 49–53.

Harvey, D. (2007). *A brief history of neoliberalism*, Oxford University Press, Oxford.

Hills, J., Sefton, T. & Stewart, K. (2009). *Towards a more equal society*, Policy Press, Bristol.

Ikonen, K-M. (2015). Precarious work, entrepreneurial mindset and the sense of place: female strategies in insecure labour markets, in Johnson, M. (Ed.). *Precariat: labour, work and politics*, Routledge, London, pp. 83–97.

James, P. (ed.). (2006) *Globalism, nationalism, tribalism: bringing theory back in*, Sage, New York.

Johnson, M. (2015). Introduction, the precariat, in Johnson, M. (Ed.). *Precariat: labour, work and politics*, Routledge, London, pp. 1–3.

Kessler, S. (2017). *Gigged: the end of jobs and the future of work*, Random House Business, New York.

Lazzarato, M. (2012). *The making of indebted man: an essay of the neoliberal condition*, Semiotext E, San Francisco.

Lele, C.G. (2011). *Organizational democracy: collaborative team culture: key to organizational growth*, Atlantic Publisher, New York.

Luhmann, N. (1996). *Social systems*, Stanford University Press, Stanford.

Mason, P. (2012). *Why it's kicking off everywhere: the new global revolutions*, Verso, New York.

Mason, P. (2015). *Postcapitalism: a guide to our future*, Allen Lane, London.

McAfee, A. & Brynjolfsson, E. (2017). *Machine platform: harnessing the digital revolution*, W.W. Norton Company, New York.

McGill, K. (2016). *Global inequality*, University of Toronto Press, Toronto.

Meister, J.C. & Mulcahy, K.J. (2017). *The future workplace experience*, McGraw Hill, New York.

Morris, J. (2012). Unruly entrepreneurs: Russian worker responses to insecure formal employment, *Global Labour Journal*, 3, 2: 217–236.

Nordberg, K. (2016). *Revolutionizing economic and democratic systems: reinventing the third way*, Palgrave, New York.

Pearson, T. (2015). *The end of jobs*, Lioncrest Publishing, New York.

Perlin, R. (2011). *Intern nation: how to earn nothing and learn little in the brave new economy*, Verso, London.

Petras, J. & Veltmeyr, H. (2011). *Beyond neoliberalism: a word to win*, Routledge, London.

Petras, J., Veltmeyr, H. & Marquez, H. (2013). *Imperialism and capitalism in the twenty-first century: a system of crises*, Routledge, London.
Piketty, T. (2014). *Capital in the twenty-first century*, Belknap, Boston.
Piketty, T. (2016). *Chronicles: on our troubled times*, Viking, London.
Pilger, J. (2016). *The new rulers of the world*, Verso, London.
Reich, R. (2015). *Saving capitalism*, Alfred. A. Knopf, New York.
Reinmoell, S. & Reinmoeller, P. (2015). *The ambidextrous organization*, Routledge, Oxford.
Roat, H. (2016). *Capital and collusion: the political logic of global economic development*, Princeton University Press, Princeton.
Rodrik, D. (2011). *The globalization paradox*, Oxford University Press, Oxford.
Rojecki, A. (2016). *America and the politics of insecurity*, John Hopkins University Press, New York.
Rosa, H. & Mathys-Treio, J. (2015). *Social acceleration*, Columbia University Press, New York.
Rozenblit, B. (2008). *Us against them. How tribalism affects the way we think*, Transcendent Publication, New York.
Savage, M. (2015). *Social class in the 21st century*, Penguin, London.
Schwab, K. (2016). *The fourth industrial revolution*, World Economic Forum, Geneva.
Sennett, R. (2003). *The fall of public man*, Penguin, New York.
Sennett, R. (2013). *Together*, Penguin, New York.
Susskind, R. & Susskind, D. (2015). *The future of professions: how technology will transform the work of human experts*, Oxford University Press, Oxford.
Tarrow, S. (2005). *The new transnational activism*, Cambridge University Press, Cambridge.
Tolonen, T. (2005). Locality and gendered capital of working-class youth, *Young*, 13, 4: 343–361.
Trot, B. (2015). From the precariat to the multitude, in Johnson, M. (Ed.). *Precariat: labour, work and politics*, Routledge, London, pp. 22–41.
Varga, J.J. (2015). Breaking the heartland. Creating the precariat in the US lower rust belt, in Johnson, M. (Ed.). *Precariat: labour, work and politics*, Routledge, London, pp. 46–62.
Wakeling, P. & Savage, M. (2015a). Elite universities, elite schooling and reputation in Britain, in Zanten, A.V. & Ball, S. *Elites, privilege and excellence: the national and global redefinition of educational advantage, world yearbook of education*, Abingdon, London.
Wakeling, P. & Savage, M. (2015b). Entry to elite positions and the stratification of higher education in Britain, *Sociological Review*, 63, 2: 290–320.
Wiedemer, D., Wiedemer, R.A. & Spitzer, C. S. (2015). *Aftershock*, Wiley, London.
Wilson, M. (2017). *Implementation of robot systems*, Butterworth-Heinemann, New York.
Zhao, J., Feng, Z., Chu, F. & Ma, N. (2017). *Advanced theory of constraint and motion analysis for robot mechanisms*, Academic Press, London.

5 Concepts

Ambidextrous organizations

Ambidextrous organizations are organizations that have the ability to adapt to changes in external conditions while at the same time generating their own future by means of, among other things, performance improvement, growth and innovation (Duncan, 1976; O'Reilly & Tushman, 2004; 2007; 2011; Thota & Munir, 2011). In Chapter X, we have shown how ambidextrous organizations can be developed by HR departments.

In 2004, O'Reilly & Tushman expressed that ambidextrous organizations would constitute one of the major challenges for management in the global knowledge economy.

The findings of O'Reilly & Tushman (2004) were overwhelming. Regarding the launching of radical innovations, they found that none of the cross-functional or unsupported teams and only a quarter of the teams with functional designs were able to produce radical innovations. However, among the ambidextrous organizations, 90% were successful in producing radical innovations. Empirical research has shown that this type of organizational design is best for producing both incremental and radical innovations.

Asplund's motivation theory[1]

In brief, this theory can be described in the following way: *People are motivated by social responses* (Asplund, 2010: 221–229). The following statement may be said to be a central point made by Asplund's theory: *When people receive social responses, their level of activity increases.*

Asplund's motivation theory is consistent with North's action theory (ref. North's action theory). Understood in this way, it seems reasonable to connect the two theories in the statement: *People are motivated by the social responses rewarded by the institutional framework.*

Availability cascades

This refers to the idea that we are all controlled by the image of reality created by the media, because this image is easy to retrieve from memory.

Availability proposition

This may be expressed as follow: the more easily information enters into our consciousness, the greater the likelihood that we will have confidence in that information. In other words, we believe more in the type of information that is available in memory than the information that is not so readily available.

Behavioural perspective

This perspective focuses on the behaviour of employees as an explanation for the relationship between business strategy and the results obtained.

Boudon-Coleman diagram

This research methodology was developed by Mario Bunge (Bunge, 1977: 76–79) based on insights made by the sociologists Boudon and Coleman. The purpose of the diagram is to show the relationship between the various levels, such as the macro- and micro-levels. For instance, it is shown how changes at the macro-level, such as technological innovations in feudal society, can lead to increased income at the micro-level. However, it was shown that technological innovations could lead to weakening of the semi-feudal structures because dependency on landowners was reduced. Consequently, the landowners opposed such changes especially in the case of technological innovations, which Boudon has shown in his research (Boudon, 1981: 100). Coleman (Coleman, 1990: 7–12) started at the macro-level, went to the individual level to find explanations and finally ended up at the macro-level again.

An important purpose of Bunge's Boudon-Coleman diagram is to identify social mechanisms that maintain or change the phenomenon or problem under investigation (as mentioned above, in Boudon's analysis of semi-feudal society). Bunge's Boudon-Coleman diagram may be said to represent a "mixed strategy"; Bunge says the following:

> When studying systems of any kind a) reduce them to their components (at some level) and the interaction among these, as well as among them and environmental items, but acknowledge and explain emergence (see the chapter on concepts) whenever it occurs, and b) approach systems from all pertinent sides and on all relevant levels, integrating theories or even research fields whenever unidisciplinarity proves to be insufficient.(Bunge, 1998: 78)

The purpose of this research strategy is to arrive at a deeper and more complete explanation of a system's behaviour.

Capabilities

Capabilities are for an organization what abilities are for an individual.

An organizational capability may thus be defined as an organization's ability to perform a task, activity or process. Operational capabilities enable an organization to make money in the here and now (Winter, 2003: 991–995). Dynamic capabilities, as opposed to operational capabilities, are linked to processes of change. Change and innovation are at the centre of dynamic capabilities.

Simplified, one may say that organizational capabilities are something an organization does well compared to its competitors (Ulrich and Brockbank, 2005). These capabilities are intangible and therefore difficult for competitors to imitate (Wernerfelt, 1984).

Cohesive energy

In a social system cohesive energy is "the glue" that binds the system together. Cohesive energy is the social mechanisms that make the system durable. According to systemic thinking it is the relationships and actions that bind social systems together. The rationale is that relationships and the systems of relationships may be said to control human behaviour. Social systems are held together (in systemic thinking) by dynamic social relations (e.g. feelings, perceptions, norms) and social action (e.g. cooperation, solidarity, conflict and communication).

Co-creation

Co-creation involves working together to promote knowledge processes and innovation. If knowledge processes and innovation are essential for value creation in the knowledge society, co-creation is an important social mechanism for initiating, maintaining and strengthening these processes. The balance between competition and cooperation, embodied in the concept of co-creation, leads to constructive criticism and the necessary scope of knowledge that exists in the network so as to promote creativity and the innovative. Instead of a zero-sum situation, a positive-sum situation will be developed where everyone wins.

Collective blindness

Collective blindness may be said to be a form of collective arrogance, which results in irrational actions. Minor events slip under the radar, causing the system to not be fully aware of what is happening. Politicians' explanations why voters in a referendum vote contrary to what most of the power elite and the media advocated is an example of collective blindness.

Competence

Competence refers to knowledge, skills and attitudes.

Core competence

The concept was popular in the strategy literature of the 1990s. Core competence may be defined as: "a bundle of skills and technologies that enable a company to provide a particular benefit to customers" (Hamel & Prahalad, 1996: 219). More recently, core competence as a concept has been given less attention in the research on dynamic capabilities, and now there is more focus on the concept of *fitness*. The term *evolutionary fitness* is also used in the research literature in connection with technology, quality, cost development, market development, innovation and competitive positioning (Helfat et al., 2007: 7).

Discontinuous innovations

These are innovations that change the premises of technology, markets, our mindset and so on. We know that sooner or later discontinuous innovations will emerge in the future (Hewing, 2013).

Dynamic capabilities

Dynamic capabilities stem from the resource-based perspective and evolutionary thinking in strategy literature (Teece, 2013: 3–65, 82–113; Nelson and Winter, 1982). The dynamic perspective attempts to explain what promotes an organization's competitive position over time through innovation and growth (Teece, 2013: x).

The original thinking concerning dynamic capabilities may be related to Teece et al. (1997). These authors defined dynamic capabilities as *an organization's ability to create, develop and modify its internal and external expertise in order to address changes in the external world.*

Dynamic capabilities are now seen as all the organizational processes, not only internal and external expertise, that contribute to an organization's capacity to adapt to change while creating the organization's future.

Explicit knowledge

This is knowledge that can be digitized and communicated to others as information.

Evidence

This may be results, such as research results, that can be relied on. However, it is also important to be aware of the fact that other evidence may be available without having to refer to figures and quantities, such as evidence that emerges from observations and good judgement without the assessment being quantified. Evidence-based research is research results that are based on approved and accepted scientific research methods.

Emergent

An emergent occurs if something new turns up on one level that has not previously existed on the level below. With emergent we mean: Let S be a system with composition A, i.e. the various components in addition to the way they are composed. If P is a property of S, P is emergent with regard to A, if and only if no components in A possess P; otherwise P is to be regarded as a resulting property with regards to A (Bunge, 1977: 97).

Entrepreneurial spirit

The entrepreneurial spirit may be described as follows (Roddick, 2003: 106–107):

- The vision of something new and belief in this that is so strong that belief becomes reality.
- A touch of positive madness.
- The ability to stand out from the crowd.
- Creative tension bubbling over.
- Pathological optimism.
- To act before you know!
- Basic desire for change.
- Creative energy focused on ideas, not on explicit factual knowledge.
- Being able to tell the story you want to sell.

Feedback

Giving the other person feedback, for instance with regard to their behaviour, attitudes and the like, is the most important element in the area of interactive skills and emotional intelligence (Goleman, 1996; 2007). Analysis of feedback is a sure way to identify our strengths and then reinforce them (Wang et al., 2003). Failure to give people feedback on their behaviour in some contexts may even be considered immoral.

Feed-forward

Feed-forward is regarded here as an expectation mechanism. It seems reasonable to assume that our expectations influence our behaviour in the present. It is therefore important that we make explicit to ourselves the expectations we have of a situation. By making expectations explicit, we have a greater opportunity to learn from our experiences and thus improve our performance.

Front-line focus

This refers to those in the front line, i.e. in direct contact with customers, users, patients, students, etc. They have the greatest expertise, necessary information and decision-making authority and are regarded as the most important resource in the organization because they are at the point where an organization's value creation occurs.

Global competence network

These competence networks may be divided into political, social, economic, technological and cultural patterns. It is when these five patterns interact that one may perceive the overall pattern. In the global knowledge economy it seems reasonable to assume that those who control this pattern set the conditions for economic development. These global competence networks will most likely make an impact on HR departments in companies competing for this kind of expertise in national markets.

Global competence networks are also emphasized as crucial for economic growth by OECD (2001), although they use the term *innovative clusters.* The purpose of innovative clusters and global competence networks is the development, dissemination and use of new ideas that promote wealth creation.

There is much to suggest that a greater degree of integration and cooperation between private and public sectors at the national and regional levels is an important prerequisite for initiating the innovative locomotive effect. The global competence networks are metaphorically the energy source that sustains the motion of this locomotive. It would be counterproductive to replace the locomotive once in motion. Conversely, the individual carriages of the locomotive (read: organizational level) can be changed depending on their competitive position. The individual passengers on the train create ideas and knowledge through the processes that may be called *creative chaos.* In this way we will arrive at a tripartite of the prerequisites for global competence networks. At the individual level, creative chaos occurs. At the organizational level, there will be creative destruction. At the social and global levels, creative collaboration takes place. These three processes create innovation and economic growth as an emergent, not as a *future perfectum*, i.e. a planned process with given results.

A prerequisite for the reasoning above is that tension and competition at one level requires collaboration at another level. Competition and cooperation are both necessary if one is to develop innovation and economic growth, in the same manner that stability and change are necessary for flexibility. Too much of the one (stability) leads to rigidity, and too much of the other (change) leads to chaos. Understood in this way, emergents cannot be planned.

Hamel's Law of Innovation

The "law" states that only between one and two of one thousand ideas become innovations in a market (Hamel, 2002; 2012). Therefore, an infostructure must be created to ensure that ideas are continuously produced in a business.

Hidden knowledge

Hidden knowledge is what we do not know we do not know. Kirzner (1982) says that hidden knowledge is possibly the most important knowledge domain of creativity, innovation and entrepreneurship.

History's "slow fields"

This refers to the fact that norms, values and actions tend to be in operation long after the functions, activities and processes that initially created them disappear, thus generating so-called *slow fields of history*. These norms, values and actions exist though they have no apparent function, contributing to maintaining a type of behaviour long after the type of behaviour is functional or meaningful.[2] For sociologists and historians it is important to determine whether norms and values have any function, or whether they are part of history's slow fields. By examining history's slow fields, it may be possible to provide better explanations for phenomena.

HR management

HR management is defined as HR practices at various levels (micro, meso, macro) for managing people in organizations.

HR management has been defined in many different ways. For instance, Boxall and Purcell (2003: 1) define HR management as all those activities oriented towards managing relations between employees in an organization. This definition emphasizes the relational perspective. Later, they expanded their definition to include all the activities and processes that underpin an organization's value creation (Boxall and Purcell, 2010: 29). On this basis, Armstrong defines the activities and processes that HR management should engage in:

> HRM covers activities such as human capital management, knowledge management, organizational design and development, resource planning (recruitment, talent development), performance management, organizational learning, reward systems, relationships between employees, and employees' wellness.(Armstrong, 2014: 6)

However, we believe Armstrong underestimates two essential areas of knowledge in his definition: the management of innovation processes, and change processes in organizations. Innovation and change are strongly emphasized in the global HRM survey (White & Younger, 2013: 35–39). Armstrong has included the ethical perspective in his handbook for HRM (Armstrong, 2014: 95–105). Management of innovation processes and change processes in organizations is also highlighted and underlined by Wright et al. (2011: 5) in their description of HRM. However, it must also be said that Armstrong discusses innovation (Armstrong, 2014: 145–155), but not in his process definition of HR management. Innovation and change processes are also emphasized by Ulrich et al. (2013a). Brockbank (2013: 24) especially mentions these two processes as being important in the research model Ulrich et al. (2013a) have developed through their empirical research over 25 years.

Implicit knowledge

This is knowledge that is spread throughout an organization but not integrated.

Informats

By informat, we mean robots with artificial intelligence which are interconnected in a global technological network. Figuratively, informats may be imagined as clusters of neurons in the human brain, which are connected to other neuron-clusters to create the various functions of human intelligence (Wilson, 2017; Winfield, 2012; Vadakkepat & Goswami, 2018). In the financial world, the use of robots by financial analysts provides an example of the use of so-called informats. Informats for use in medical surgery are already on the drawing board and will be a reality in the near future (Bleuer et al., 2017); the same applies to the use of informats in the service and education sectors (Bleuer & Bouri, 2017). Informats are understood here in the context of the above description as being emergents[3] in relation to robots; informats can sense, analyse and reach decisions in the space of a micro-second.

Information input overload

This occurs when an individual, a team, an organization or a community receive more information than they can manage to process.

In a situation characterised by information input overload the following may occur (Miller, 1978: 123):

1. Designated tasks and responsibilities are left undone.
2. Errors are made.
3. Queues of information occur.
4. Information is filtered out that should have been included.
5. Abstract formulations are made when they should have been specific.
6. Communication channels are overloaded, creating stress and tension in the system.
7. Complex situations are shunned.
8. Information is lumped together for processing.

Each of the above eight points may result in a decrease in efficiency when the system is exposed to information input overload.

Informats

By informat, we mean robots with artificial intelligence which are interconnected in a global technological network. Figuratively, informats may be imagined as clusters of neurons in the human brain, which are connected to other neuron-clusters to create the various functions of human intelligence (Wilson, 2017; Winfield, 2012; Vadakkepat & Goswami, 2018). In the financial world, the use of robots by financial analysts provides an example of the use of so-called informats. Informats for use in medical surgery are already on the drawing board and will be a reality in the near future (Bleuer et al., 2017); the same applies to the use of informats in the service and education sectors (Bleuer & Bouri, 2017). Informats are understood here in the context of the above description as being emergents[4] in relation to robots; informats can sense, analyse and reach decisions in the space of a micro-second.

Infostructure

The infostructure concerns the processes that enable the development, transfer, analysis, storage, coordination and management of data, information and knowledge. The infostructure consists of eleven generic processes. The eleven processes in the infostructure may be considered as nodes in a social network at different levels, for example team, organization, society and region, all in the global space. Together, the eleven processes comprise the totality of the infostructure.

It may be said that the *info*structure has the same importance in the knowledge society as the *infra*structure had in the industrial society.

Innovation

Innovation is here understood as any idea, practice or material element, which is perceived as new for the person using it (Zaltman et al., 1973).

Ideas are seen as the smallest unit in the innovation process (Hamel, 2002; 2012). However, this refers to the ideas that are in process of development and not fully developed ideas. Before an idea can be characterized as innovative, it must prove to be beneficial to somebody, i.e. the market must accept the idea and apply it. Consequently, the creative process of innovation is here understood as the benefit it has for a market. Thus, it is not sufficient that an idea is new for it to be considered an innovation. An idea may have a great degree of novelty, but if it is of no benefit to anybody in the market, then it has no innovative value.

Kaizen

This is a Japanese method, which means that an organization develops systems for organized improvement.

Knowledge

The definition of knowledge used here is *the systematization and structuring of information for one or more goals or purposes.*

Knowledge enterprise

This is an enterprise that has knowledge as its most significant output. It is perhaps helpful to think of the process *input–process–output* to separate industrial enterprises from knowledge enterprises. Much knowledge and skills are needed to produce high-tech products such as computers, and there are also many knowledge workers involved in this process. However, the majority of products produced today are high-tech industrial products, and although such products require very skilled knowledge in the production process, they are nevertheless output-industrial products.

On the other hand, law firms, consulting firms and universities are examples of knowledge enterprises.

Knowledge management

Management of knowledge resources in an organization. These resources may be explicit knowledge, implicit knowledge, tacit knowledge and hidden knowledge.

Knowledge worker

A knowledge worker has been described by the OECD as *a person whose primary task is to generate and apply knowledge*, rather than to provide services or produce physical products (OECD, 2000a, 2000b, 2000c, 2000d, 2000e; 2001). This may be understood as a *formal definition* of a knowledge worker.

This definition does not restrict knowledge workers to creative fields, as is the case with, for example, Mosco and McKercher (2007: vii–xxiv). The OECD definition also allows for the fact that a knowledge worker may perform routine tasks. The definition also does not limit the type of work performed by knowledge workers to tasks relating to creative problem-solving strategies, unlike the definition provided by Reinhardt et al. (2011).

Locomotive effect

This refers to something that generates and then reinforces an activity or development.

Modularization

An extreme fragmentation of the production process in the global knowledge economy. Production is fragmented and distributed according to the following logic: costs–quality–competence–design–innovation.

Modular flexibility

The modulization of value creation. Modular flexibility may best be understood as the globalization of production processes, and extreme specialization of work processes with a focus on core processes.

Necessary and sufficient conditions

It may often be appropriate to divide conditions or premises into *necessary conditions* and *sufficient conditions.* Necessary conditions must be present to trigger an action, but these may not be sufficient. The sufficient conditions must also be present to trigger the action.

North's action theory[5]

This action theory may be expressed in the following statement: People act on the basis of a system of rewards as expressed in the norms, values, rules and attitudes in the culture (the institutional framework) (North, 1990; 1993). North's action theory is also consistent with Asplund's motivation theory (ref. Asplund's motivation theory).

Primary task

An organization's primary task is what the system is designed to do.

Proposition

This is an overarching hypothesis. It says something about the relationship between several variables. A proposition relates to a hypothesis in the same way the main research problem relates to research questions.

Punctuation

By punctuation (Bateson, 1972: 292–293) a distinction is drawn between cause and effect; this is done with a clear motive in mind. A causality is thus created which does not actually exist in the real world, and one is then free to discuss the effects of this cause which has been created through a process of punctuation.

A sequence of a process is selected, and then bracketed. In this way, we de-limit what is punctuated from the rest of the process. Figuratively, we may imagine this as a circle that is divided into small pieces; one piece of the circle is then selected and folded out into a straight line. This results in the creation of an artificial beginning and end. This beginning and end of course cannot exist in a circle, but only through the process of punctuation.

Social laws

Social laws constitute a pattern of a unique type. They are systemic and connected to a system of knowledge, and cannot change without the facts they represent also being changed (Bunge, 1983a; 1983b). The main differences between a statement of a law and other statements are:

1 Law statements are general.
2 Law statements are systemic, i.e. they are related to the established system of knowledge.
3 Law statements have been verified through many studies.

A pattern may be understood as variables that are stable over a specific period of time. A social law is created when an observer gains insight into the pattern. By gaining such insight, we can also predict parts of behaviour or at least develop a rough estimate within a short period of time.

Social laws are further related to specific social systems, both in time and space. However, this does not represent any objection to social laws, because this is also true of natural laws (although these have a longer time span and are of a more general nature).

Social mechanism

Robert Merton (1967) brought the notion of social mechanisms into sociology, although we can find rudiments of this in both Weber – with the Protestant ethic as an explanation for the emergence of capitalism in Europe – and in Durkheim, who uses society as an explanation for a rising suicide rate. For Merton, social mechanisms are the building blocks of *middle range theories.* He defines social mechanisms as *social processes having designated consequences for designated parts of the social structure* (Merton, 1967: 43). In the 1980s and 1990s, Jon Elster developed a new notion of the role of social mechanisms in sociology (Elster, 1986; 1989). Hedstrom and Swedberg write that "the advancement of social theory calls for an analytical approach that systematically seeks to explicate the social mechanisms that generate and explain observed associations between events" (Hedstrøm & Swedberg, 1998 : 1).

It is one thing to point out connections between phenomena. It is something quite different to point out satisfactory explanations for these relationships, which is what social mechanisms accomplish. A social mechanism tells us what will happen, how it will happen and why it will happen (Bunge, 1967). Social mechanisms are primarily analytical constructs which cannot necessarily be observed; in other words, they are epistemological, not ontological. However, social mechanisms are observable in their consequences. An intention can be a social mechanism of action. We cannot observe an intention, but we can interpret it in light of the consequences manifested through an action. Preferences can also function as a social mechanism for economic behaviour. We cannot observe a person's preferences, but we can interpret them in the light of the behavioural consequences that manifest themselves. Social mechanisms are, understood in this way, analytical constructs, indicating connections between events (Hernes, 1998).

Bunge says: "a social mechanism is a process in a concrete system, such that it is capable of being about or preventing some change in the system as a whole or in some of its subsystems" (Bunge, 1997: 414). By social mechanism here we mean those activities that promote/inhibit social processes in relation to a specific problem/phenomenon.

Material resources and technology are social mechanisms of the economic subsystem; power is a social mechanism of the political subsystem; fundamental norms and values are a social mechanism of the cultural subsystem; and human relationships are a social mechanism of the social subsystem. These system-specific social mechanisms interact with each other to achieve certain goals, to maintain these systems or to avoid certain undesirable conditions in the system or the outside world.

The difficulty of discovering social mechanisms and distinguishing them from processes may be partly explained by the fact that social mechanisms are also processes (Bunge, 1997: 414). For the application of social mechanisms, see the Boudon-Coleman diagram.

Social system

From a systemic perspective, social systems can be conceptual or concrete. Theories and analytical models are examples of conceptual systems. Further, social systems are *composed of people and their artifacts* (Bunge, 1996: 21). Social systems are held together (in systemic reasoning) by *dynamic social relations* (such as emotions, interpretations, norms, etc.) and *social actions* (such as cooperation, solidarity, conflict and communication, etc.). None of the social actions have precedence in the systemic interpretation of social systems, such as conflict in the case of Marx, and solidarity in the case of Durkheim.

Staccato-behaviour (erratic behaviour)

If organizations introduce too many change processes in succession too quickly, a phenomenon may occur called "staccato-behaviour".

If an organization does not deal with this appropriately, it seems reasonable to assume that workers will become tired, burnt-out and de-motivated. Perhaps most damaging to business, employees will lose focus on their primary task – what the business is designed to do. In addition, businesses will often experience that this leads to an increasing degree of opportunistic behaviour (Ulrich, 2013b: 260).

Strategic HR management

Strategic HR management is defined in this book as: *The choices an HR department makes with regard to human resources for the purposes of achieving the organization's goals.* This is analogous to the view of Storey et al. (2009: 3) and consistent with the definition we employ of HR management. This means that strategic HR management must be focused on the *micro, meso* and *macro-levels.*

There are many definitions of strategic HR management. For instance, use of human resources in order to achieve lasting competitive advantages for the business; management of the employees, expressed through management philosophy, policy and praxis; development of a consistent practices in order to support the strategic goals of the business; a complex system with the following characteristics: vertical integration, horizontal integration, efficiency, partnership.

Systemic thinking

Systemic thinking makes a distinction between the epistemological sphere (Bunge, 1985), the ontological sphere (Bunge, 1983a), the axiological sphere (Bunge, 1989, 1996) and the ethical sphere (Bunge, 1989). Systemic thinking makes a clear distinction between intention and behaviour. Intention is something that should be *understood*, while behaviour is something that should be *explained.* To understand an intention we must study the historical factors, situations and contexts, as well as the expectation mechanisms. Behaviour must be explained with respect to the context, relationships and situation it unfolds in. What implication does the distinction between intention and behaviour have for the study of social systems?

Interpretation of meaning is an important part of the *intention aspect* in the distinction. Explanation and prediction become an essential part of the *behavioural aspect* of the distinction.

In systemic thinking it is the link between the interpretation of meaning and explanation, and prediction, which provides historical and social sciences with practical strength. By making a distinction between intention and behaviour, the historical and the social sciences are interpretive, explanatory and predictive projects. According to systemic thinking, many of the contradictions in the historical and social sciences spring from the fact that a distinction is not made between intention and behaviour. The problem of the historical and social sciences is that the actors who are studied have both intentions and they also exercise types of behaviour; however, this isn't problematic as long as we make a distinction between intention and behaviour. By simultaneously introducing the distinction between intention and behaviour, systemic thinking has made it possible to identify, for instance, partial explanations from each of two main epistemological positions, namely, the naturalists and anti-naturalists (Johannessen & Olaisen, 2005; 2006), and synthesize these explanations into new knowledge.

Systemic thinking emphasizes circular causal processes, also called *interactive causal processes*, in addition to linear causal processes (Johannessen, 1996; 1997). Systemic thinking argues that to understand objective social facts, one must examine the subjective aspects of these. In systemic thinking, objective social facts exist, but they are often more difficult to grasp than facts in the natural world, because social facts are often influenced by expectations, emotions, prejudices, ideology and economic and social interests. *"Aspect-seeing"* is thus a way of approaching these social facts.

Emergents are central to systemic thinking. A pattern behind the problem or phenomenon is always sought in systemic investigations. Patterns may be revealed by studying the underlying processes that constitute a phenomenon or problem, *and the search for pattern is what scientific research is all about* (Bunge, 1996:42).

According to systemic thinking it is a misconception to say that the facts are social constructions. The misunderstanding involves confusing our *concepts* concerning facts and our *hypotheses* about the facts together with the facts. Our concepts and hypotheses are mental constructs. The facts, however, are not mental constructs. Social need, for instance, is not a social fact; it is a mental construct of, for instance, starvation. Starvation is a social fact. Social need is a mental or social construction. Not being able to read is a social fact. Illiteracy is, however, a social construction.

A *symbol* should symbolize something, just as a *concept* should delineate something. A *hypothesis* should explain something or express something about relationships. A conceptual *model* should say something about the relationships between concepts. A *theory* should say something about relationships between propositions. Physical or social facts are untouched by all these mental constructions. That one can through constructs change social facts, or that social facts are changed as a social consequence of using constructs, is neither original nor new.

The aim of theoretical research, according to the systemic position, is the construction of systems, i.e. theories (Bunge, 1974: v). The order in systemic research is thus: Theory–Analysis–Synthesis.

In the methodological sphere, the systemic position has its main focus on relationships, both in terms of concrete things, ideas and knowledge. Consequently, systemic thinking encourages interdisciplinary and multidisciplinary approaches to problems or phenomena.

The systemic position thus attempts to bridge the gap between methodological individualism and methodological collectivism, which is considered the classic controversy in historical and social sciences.

The perceptions that an observer has about social systems will influence his/her actions, regardless of whether the perceptions are true or fallacious. Systemic investigations start, therefore, writes Bunge *from individuals embedded in a society that preexists them and watch how their actions affect society and alter it* (Bunge, 1996: 241). The study of social systems from a systemic perspective for these reasons always includes the triad: actors–observers–social systems.

The observer tries to uncover a system's composition, environment and structure. Then the actors' subjective perception of composition, environment and structure are examined. In other words, both the subjective and objective aspects are studied. When we wish to study changes in social systems, from a systemic point of view, we have to examine the social mechanisms (drivers) that influence changes; both internal and external social mechanisms must be identified. This study takes place within the four subsystems: the economic, political, cultural and relational. According to systemic thinking, social changes occur along seven axes:

1 As an *expectation* of new relationships, values, power constellations, technologies and distribution of material resources.
2 As a result of our *beliefs* (mental models) about relationships, values, power constellations, technical and material resources.
3 As a result of *psychological elements*, such as: irritation, crisis, discomfort, unsatisfactory life, unworthy life, loss of well-being, etc.
4 As a result of *communication* in and between systems.
5 As a result of an *understanding of connections* (contextual understanding).
6 As a result of learning and new *self-knowledge.*
7 As a result of *new ideas* and ways of thinking.

Historiography, from a systemic perspective, has one clear goal: to investigate what happened, where it happened, when it happened, how it happened, why it happened, and with what results.

Systemic assumptions related to historiography and social sciences may be expressed in the following (Bunge 1998: 263):

a The past has existed.
b Parts of the past can be known.
c Every uncovering of the past will be incomplete.
d New data, techniques, and systemizations and structuring will reveal new aspects of the past.
e Historical knowledge is developed through new data, discoveries, hypotheses and approaches.

In systemic thinking if changes are to take place, then the material will sometimes be given precedence; at other times, ideology, ideas and thinking are given precedence. In other contexts, there is a systemic link between the material and ideas that is needed to bring about changes. In such contexts, it is difficult and irrelevant to say what is the primary driver, i.e. the material or ideas; this would be on par with discussing what came first, the chicken or the egg.

The processes that drive social change, according to a systemic perspective, are the interaction between the economic, political, relational and cultural subsystems. In some situations, one of these four perspectives will prevail, whereas in others it will be one or more of the four subsystems that are the drivers of social change. In many cases, it is precisely the interaction between the four subsystems that leads to social changes.

In this context the systemic perspective may be described by saying that material conditions/energy, such as economic relationships, may provide the ground from which ideologies develop, but that these ideologies in return influence the development of the material. Whether material conditions/energy or ideology comes first is often determined by a historiographical punctuation process (Bateson, 1972: 163).

The systemic perspective balances historical materialism and historical idealism. It assumes that overall social changes are the result of economic, political, social and cultural factors, in addition to the interaction between material conditions/energy and ideas. Furthermore, a systemic perspective views any society as being interwoven into its surroundings (Bunge, 1998: 275). When a historian considers a historical situation – such as the massacre in Van in April 1915 – from this perspective then he is trying to throw light upon the internal working of a past culture and society (Stone, 1979: 19).

The systemic position attempts to view the relevant event in a larger context, in order to find the patterns which combine (Bateson, 1972: 273–274), because change depends upon feedback loop (Bateson, 1972: 274). Bunge says about this position: "By placing the particular in a sequence, adopting a broad perspective the systemist overcomes the idiographic/nomothetic duality … as well as the concomitant narrative/structural opposition" (Bunge 1998: 275). This means, metaphorically, that the systemic researcher uses a microscope, telescope and a helicopter to investigate patterns over time.

Systemic research strategy is a zigzagging between the micro-meso and macro levels (Bunge, 1998: 277). Through a systemic research strategy the researcher has ample opportunities to use a Boudon-Coleman diagram.

Systemic thinking examines four types of changes:[6]

- Type I change concerns individuals who change history, such as Genghis Khan, Hitler, Stalin, Mao Zedong, etc.
- Type II change concerns groups of people acting together who change history. Examples of Type II change include the invasion of the Roman Empire by peoples from the north, and the Ottoman expansion into the Balkans between the late 1400s and when the Ottoman Empire was pushed back partly due to nationalist liberation movements in the early 1900s.
- Type III change include changes in history that are caused by natural disasters, such as the volcanic eruption that destroyed Pompeii. Climate change may also be said to an example of a type III change.
- Type IV change involves a total change in the way of thinking, such as the emergence of new religions, like Islam, or a new political ideology, such as Marxism.

The systemic researcher attempts to explore the relationship between the four types of changes. A single event is in itself not necessarily of special interest to the systemic researcher; rather, the focus is on the *system of events* of which the single event is a part.

All the social sciences are used in the systemic position to seek insight, understanding and to explain a phenomenon or problem.

Tacit knowledge

Knowledge that is difficult to communicate to others as information. It is also very difficult, if at all possible, to digitize.

The knowledge-based perspective

The knowledge-based perspective is defined here as creating, expanding and modifying internal and external competencies to promote what the organization is designed to do (Grant, 2003: 203).

The resource-based perspective

This perspective can be defined as the structuring and systematization of the organization's internal *resources* so it is difficult for competitors to copy them.

Theory

Here understood as a system of propositions (Bunge, 1974: v).

Notes

1 Asplund's motivation theory, a term we use here, is based on Asplund's research.
2 Asplund (1970: 55) refers to a similar phenomenon when he discusses Simmel. He points out that the norms that may have had a positive function during a historic phase become in a later phase dysfunctional.
3 Emergents appear if something new occurs on a level that did not previously exist on the level below. By emergent we mean: "Let S be a system with composition A, i.e. the various components in addition to the way they are composed. If P is a property of S, P is emergent with regard to A, if and only if no components in A possess P; otherwise P is to be regarded as a resulting property with regards to A" (Bunge, 1977: 97).
4 Emergents appear if something new occurs on a level that did not previously exist on the level below. By emergent we mean: "Let S be a system with composition A, i.e. the various components in addition to the way they are composed. If P is a property of S, P is emergent with regard to A, if and only if no components in A possess P; otherwise P is to be regarded as a resulting property with regards to A" (Bunge, 1977: 97).
5 North's action theory is a term we use here based on North's research.
6 The four types of changes are related to Bateson's (1972: 279–309) work on different types of learning, especially those discussed in his article *Logical types of learning and communication.*

References

Armstrong, M. (2014). *Armstrong's handbook of strategic human resource management*, Kogan Page, New York.

Asplund, J. (1970). *Om undran innfør samhället*, Argos, Stockholm.

Asplund, J. (2010). *Det sociala livets elementära former*, Korpen, Stockholm

Bateson, G. (1972). *Steps to an ecology of mind*, Intertext Books, London.

Bleuer, H. & Bouri, M. (2017). *New trends in medical and service robots: assistive, surgical and educational robotics*, Springer, London.

Bleuer, H., Bouri, M. & Mandada, F.C. (2017). *New trends in medical and service robots*, Springer, London.Boudon, R. (1981). *The logic of social action*, Routledge, London.

Boxall, P.F. & Purcell, J. (2003). *Strategy and human resource management*, Palgrave Macmillan, Basingstoke.

Boxall, P.F. & Purcell, J. (2010). An HRM perspective on employee participation, in Wilkinson, A., Golan, P.J., Marchington, M. & Lewins, D. (eds.), *The Oxford handbook of participation in organizations*, Oxford University Press, Oxford, pp. 129–151.

Brockbank, W. (2013). Overview and logic, in Ulrich, D., Brockbank, W., Younger, J. & Ulrich, M. (eds.), *Global HR competencies: mastering competitive value from the outside in*, McGraw Hill, New York. S. 3–27.

Bunge, M. (1967). *Scientific research, Vol. 3, in studies of the foundations methodology and philosophy of science*, Springer Verlag, Berlin.

Bunge, M. (1974). *Sense and Reference*, Reidel, Dordrecht.

Bunge, M. (1977). *Treatise on basic philosophy. Vol. 3. Ontology I: The furniture of the world*, D. Reidel, Dordrecht, Holland.

Bunge, M. (1983a). *Exploring the world: epistemology & methodology I*, Reidel, Dordrecht.

Bunge, M. (1983b). *Understanding the world: epistemology & methodology II*, Reidel, Dordrecht.

Bunge, M. (1985). *Philos sci and technology. Part I: epistemology & methodology III*, Reidel, Dordrecht.

Bunge, M. (1989). *Ethics: the good and the right*, Reidel, Dordrecht.

Bunge, M. (1996). *Finding philosophy in social science.* Yale University Press, New Haven.

Bunge, M. (1997). Mechanism and explanation, *Philosophy of the Social Sciences* 27: 410–465.Bunge, M. (1998). *Philosophy of science: from problem to theory*, Volume one, Transaction Publishers, New Jersey.

Coleman, J.S. (1990). *Foundations of social theory*, Harvard University Press, Belknap Press, Cambridge, MA.

Duncan, R. (1976). The ambidextrual organization: designing dual structures for innovation, in Kilman, R.H., Pondy, L.R. & Slevin, D. (eds.). *The management of organization*, North Holland, New York, pp. 167–188.

Elster, J. (1986). *Rational choice*, New York University Press, New York.

Elster, J. (1989). *Nuts and bolts for the social sciences*, Cambridge University Press, Cambridge.

Goleman, D. (1996). *Emotional intelligence*, Bloomsbury Publishing, New York.

Goleman, D. (2007). *Social intelligence*, Arrow Books, New York.

Grant, R.M. (2003). The knowledge-based view of the firm, in Faulkner, D. & Campell, A. (eds.). *The Oxford Handbook of Strategy*, Oxford University Press, Oxford, pp. 203–231.

Hamel, G. (2002). *Leading the revolution: how to thrive in turbulent times by making innovation a way of life*, Harvard Business School Press, Boston.

Hamel, G. (2012). *What matters now: how to win in a world of relentless change ferocious competition, and unstoppable innovation*, John Wiley & Sons, New York.

Hamel, G. & Prahalad, C.K. (1996). *Competing for the future*, Harvard Business School Press, Boston.

Hedstrøm, P & Swedberg, S.R. (1998). Social mechanisms: An introductory essay, in Hedstrøm, P & Swedberg, R. (eds). *Social mechanisms: an analytical approach to social theory*, Cambridge University Press, Cambridge.

Helfat, C. E., Finkelstein, S., Mitchell, W., Peteraf, M.A., Singh, H., Teece, D.J. & Winter, S.G. (2007). *Dynamic capabilities: understanding strategic change in organizations*, Blackwell, Oxford.

Hernes, G. (1998). Real virtuality, in Hedstrøm, P & Swedberg (eds), *Social mechanisms: an analytical approach to social theory*, Cambridge University Press, Cambridge, pp. 74–102.

Hewing, M. (2013). *Collaboration with potential users for discontinuous innovation*, Springer Gabler, Potsdam.

Johannessen, J-A. (1996). Systemics applied to the study of organizational fields: developing systemic research strategy for organizational fields, *Kybernetes*, 25, 1: 33–51.

Johannessen, J-A. (1997). Aspects of ethics in systemic thinking, *Kybernetes*, 26, 9: 983–1001.

Johannessen, J.-A. & J. Olaisen (2005). Systemic philosophy and the philosophy of social science. Part I: Transcedence of the naturalistic and the anti-naturalistic position in the philosophy of social science, *Kybernetes* 34, 7/8: 1261–1277.

Johannessen, J.-A. & J. Olaisen (2006). Systemic philosophy and the philosophy of social science – Part II: The systemic position, *Kybernetes* 34, 9/10: 1570–1586.

Kirzner, S. (1982). The theory of entrepreneurship in economic growth, in Kent, C.A., Sexton, D.L. & Vesper, K.H. (eds). *Encyclopedia of entrepreneurship*, Prentice Hall, Englewood Cliffs, NJ.

Merton, R.K. (1967). *Social theory and social structure*, Free Press, London.

Mosco, V. & McKercher, C. (2007). Introduction: theorizing knowledge labor and the information society, *Knowledge Workers in the Information Society*, Lexington Books, Lanham.

Miller, J.G. (1978). *Living systems*, McGraw-Hill, New York.

Nelson, R.R. & Winter, S.G. (1982). *An evolutionary theory of economic change*, Harvard University Press, Cambridge, MA.

North, D.C. (1990). *Institutions, institutional change and economic performance*, Cambridge University Press, Cambridge.

North, D. (1993). Nobel lecture, www.nobelprize.org/nobel_prizes/economics/laurea tes/1993/north-lecture.html#not2, (accessed 4.5.2012).

OECD (2000a). *A new economy? The changing role of innovation and information*, OECD Publishing, Paris.

OECD (2000b) *Economic outlook*, OECD Publishing, Paris.

OECD (2000c). *Education at a glance: OECD indicators*, CERI, Paris.

OECD (2000d). ICT skills and employment, working party on the information economy, Paris, 15 November, DSTI/ICCP/IE(2000)7.

OECD (2000e). *Knowledge management in the learning society*, CERI, Paris.

OECD (2001). *Innovative clusters: driving of national innovation-systems*, OECD Publishing, Paris.

O'Reilly, C.A. & Tushman, M.L. (2004). The ambidextrous organization, *Harvard Business Review*, 82, 4:74–81.

O'Reilly, C.A. & Tushman, M.L. (2007). *Ambidexterity as a dynamic capability: Resolving the innovators dilemma*, Harvard Business School Press, Boston.

O'Reilly, C.A. & Tushman, M.L. (2011). Organizational ambidexterity in action: how managers explore and exploit, *California Management Review*, 53, 4: 5–22.

Reinhardt, W., Smith, B., Sloep, P. & Drachler, H. (2011). Knowledge worker roles and actions – results of two empirical studies, *Knowledge and Process Management*, 18, 3: 150–174.

Roddick, D.A. (2003). The grassroots entrepreneur, in Elbæk, U. *Kaospilot A-Z*, Narayana Press, Gylling.

Stone, J. (1979). The revival of narrative: reflections on a new old history, *Past and Present*, 85: 3–24.

Storey, J., Ulrich, D. & Wright, P.M. (2009). Introduction, in Storey, J., Wright, P.M. & Ulrich, D. (eds). *The Routledge companion to strategic human resource management*, Routledge, London, pp. 3–15.

Teece, D.J. (2013). *Dynamic capabilities and strategic management: organizing for innovation*, OUP, Oxford.

Teece, D., Pisano, G. and Shuen, A. (1997). Dynamic capabilities and strategic management, *Strategic Management Journal*, 18, 7: 509–533.

Thota, H. & Munir, Z. (2011). *Key concepts in innovation*, Palgrave Macmillan, London.

Tushman, M.L. & O'Reilly, C.A. (1996). Ambidextrous Organization, California Management Review, 38, 4: 8-29

Ulrich, D. (2013a). Foreword, in Ulrich, D., Brockbank, W., Younger, J. & Ulrich, M. (eds.), *Global HR competencies: mastering competitive value from the outside in*, McGraw Hill, New York, pp. v–xxi.

Ulrich, D. (2013b). Future of global HR: what's next? , in Ulrich, D., Brockbank, W., Younger, J. & Ulrich, M. (eds.), *Global HR competencies: mastering competitive value from the outside in*, McGraw Hill, New York, pp. 255–268.

Ulrich, D. & Brockbank, W. (2005). *The HR value proposition*, Harvard Business School Press, Boston, MA.

Vadakkepat, P. & Goswami, P. (eds). (2018). *Humanoid robotics: a reference*, Springer, London.

Wang, Q-G., Lee, T.H. & Lin, C. (2003). *Relay feedback: analysis, identification and control*, Springer, London.

Wernerfelt, B. (1984). A resource-based view of the firm, *Strategic Management Journal*, 5, 2: 171–180.

White, J. & Younger, J. (2013). The global perspective, in Ulrich, D., Brockbank, W., Younger, J. & Ulrich, M. (eds), *Global HR competencies: mastering competitive value from the outside in*, McGraw Hill, New York, pp. 27–53.

Wilson, M. (2017). *Implementation of robot systems*, Butterworth-Heinemann, New York.

Winfield, A. (2012). *Robotics*, OUP, Oxford.

Wright, P.M., Boudreau, J.W., Pace, D.A., Libby Sartain, E.; McKinnon, P.; Antoine, R.L. (eds) (2011). *The chief HR officer: defining the new role of human resource leaders*, Jossey-Bass, London.

Winter, S.G. (2003). Understanding dynamic capabilities, *Strategic Management Journal*, 24: 991–995.

Zaltman, G., Duncan, R. & Holbeck, J. (1973). *Innovations and organizations*, Wiley, New York.

Index

"This book is the ultimate primer on green finance and carbon risk, written by somebody who has been instrumental in shaping and designing the world of carbon risk supervision described in the book. Ideal for somebody trying to understand everything you need to know to navigate the world of green finance."

– *Jakob Thomä*, Managing Director,
2 Degrees Investing Initiative

"Ezroj demystifies carbon risk and green finance, while exposing the uninitiated to the current state of the art. The book decodes the alphabet-soup of protocols and initiatives taking place around the world, and doesn't shrink from pointing out the many challenges remaining in attaining scale and standardization."

– *Evan Mills*, Founder, Energy Associates

"Our current climate conundrum needs campaigning as much as it needs scientific analysis. Responding to it necessitates policy changes at all levels, from the most local to the most international. It requires change in exotic matters such as ecological tourism, but also in more mundane sectors such as finance. In this volume, Aaron Ezroj draws upon his experience as a private practitioner and public servant, and his exposure to US federal and state, international, and EU law to paint a most insightful, non-hysterical picture of climate finance. Of what there is and what there ought to be. In a complete yet concise style he identifies the analytical tools required by both funds and investors to kill the two birds of climate and finance with one stone. Much recommended."

– *Geert Van Calster*, Head of European and
International Law, KU Leuven

"*Carbon Risk and Green Finance* provides a useful overview of the laws, regulations, and industry actions across the globe that are pushing this growing and impactful movement to incorporate climate change into financial decision-making. It's a perfect starting point for anyone wishing to familiarize themselves with this increasingly critical issue."

– *Ethan Elkind*, Director of the Climate Program,
Berkeley Law's Center for Law,
Energy and the Environment

Carbon Risk and Green Finance

As the world plans for economic recovery following the global COVID-19 pandemic, major economies are looking to comprehensive strategies for addressing carbon risks and identifying green finance opportunities. Since Bank of England Governor Mark Carney and Michael Bloomberg began tackling climate change as a financial concern, the international financial community has been developing sophisticated analytical tools that will enable the success of comprehensive efforts to address carbon risks and identify green finance opportunities.

This timely publication offers a cutting-edge analysis of the financial aspects of climate change. It discusses the most important analytical tools, their origin, how they work, where they can go, and how they fit into a larger strategy. First, reporting frameworks can allow companies to see how well they are addressing carbon risks, in particular with respect to the recommendations of the Task Force on Climate-related Financial Disclosures. Second, by quantifying how much greenhouse gas companies emit into the atmosphere as a direct or indirect result of their operations, carbon footprint calculations can help identify carbon risks with particular companies, especially within supply chains. Third, brown taxonomies can help investors identify current carbon risks by classifying fossil fuel assets in a systematic manner. Fourth, green taxonomies can help investors identify current green finance opportunities by classifying sustainable activities in a systematic manner. Fifth, scenario analysis for assets can help investors identify future carbon risks and green finance opportunities. Finally, stress testing for liabilities can help insurers and banks address future carbon risks and better inform policymakers.

Scholars, policymakers, and business professionals will find this book informative. They will gain a comprehensive understanding of the analytical tools supporting efforts to address carbon risks and identify green finance opportunities. This will hopefully make these individuals more successful in their personal endeavors to build a more sustainable and resilient economy for future generations.

Aaron Ezroj is a graduate of UCLA School of Law and a Fulbright Scholar. He was a member of California's delegation to the United Nations Climate Change Conference and managed carbon risk programs that the United Nations called groundbreaking.

Banking, Money and International Finance

17. **Financial Risk Management in Banking**
    Evidence from Asia Pacific
    Shahsuzan Zakaria; Sardar M. N. Islam

18. **The Economics of Financial Cooperatives**
    Income Distribution, Political Economy and Regulation
    Amr Khafagy

19. **Financial Integration in the European Monetary Union**
    Sławomir Ireneusz Bukowski

20. **Banking in China 1890s–1940s**
    Business in the French Concessions
    Hubert Bonin

21. **Management of Foreign Exchange Risk**
    Evidence from Developing Economies
    Y. C. Lum and Sardar M. N. Islam

22. **Banking and Effective Capital Regulation in Practice**
    A Leadership Perspective
    Sophia Velez

23. **Innovation in Financial Services**
    Balancing Public and Private Interests
    Edited by Jan Monkiewicz and Lech Gąsiorkiewicz

24. **Carbon Risk and Green Finance**
    Aaron Ezroj

For more information about this series, please visit: www.routledge.com/series/BMIF

Carbon Risk and Green Finance

Aaron Ezroj

Routledge
Taylor & Francis Group
LONDON AND NEW YORK

First published 2021
by Routledge
2 Park Square, Milton Park, Abingdon, Oxon OX14 4RN

and by Routledge
52 Vanderbilt Avenue, New York, NY 10017

Routledge is an imprint of the Taylor & Francis Group, an informa business

British Library Cataloguing-in-Publication Data
A catalogue record for this book is available from the British Library

Library of Congress Cataloging-in-Publication Data
Names: Ezroj, Aaron, author.

Title: Carbon risk and green finance/Aaron Ezroj.
Description: Abingdon, Oxon ; New York, NY : Routledge, 2021. |
Series: Banking, money and international finance |
Includes bibliographical references and index.
Identifiers: LCCN 2020035517 (print) | LCCN 2020035518 (ebook) |
ISBN 9780367559915 (hardback) | ISBN 9781003095996 (ebook)
Subjects: LCSH: Carbon dioxide mitigation – Economic aspects. |
Finance – Environmental aspects. | Carbon offsetting.
Classification: LCC HC79.P55 E99 2021 (print) | LCC HC79.P55 (ebook) |
DDC 363.738/74561 – dc23
LC record available at https://lccn.loc.gov/2020035517
LC ebook record available at https://lccn.loc.gov/2020035518

ISBN: 978-0-367-55991-5 (hbk)
ISBN: 978-1-003-09599-6 (ebk)

Typeset in Sabon
by Apex CoVantage, LLC

I would like to dedicate this book to my parents, who taught me to be kind and work hard.

Contents

Figures

Tables

Preface

I started my legal career advising on the drafting of a California climate change regulation aimed at curbing global greenhouse gas emissions. After recently finishing UCLA School of Law and a Fulbright Scholarship, I moved to California's capital, Sacramento. There I clerked at the California Air Resources Board, where I helped the agency exercise its rulemaking authority as envisioned under the Global Warming Solutions Act.

After completing my time with the California Air Resources Board, I continued my career in public service, working as a public interest attorney throughout California. Meanwhile, I continued to keep an eye on full-time positions with the state and was eventually hired as an attorney with the California Department of Insurance. Around that time, it just so happened that the commissioner of the California Department of Insurance had ambitious plans for tackling climate risk, and after expressing interest in such efforts, I was made the director of the Office of Climate Risk Initiatives.

As the director of the Office of Climate Risk Initiatives, I worked to make climate risk a mainstream financial consideration. I managed one of the world's only databases of fossil fuel holdings. I oversaw a first-in-the-nation scenario analysis for financial assets. I also co-authored a policy paper that discussed the impact of climate change on the increasing severity of California's wildfires.

At the same time, I was able to work with an amazing range of individuals while representing my state in international United Nations conferences. This included speaking on our fossil fuel database at one of the first meetings of the Network for Greening the Financial System in the Netherlands. This involved presenting scenario analysis results to financial regulators in Brazil. Moreover, this included joining California's delegation to the United Nations Convention on Climate Change in Poland.

Because I have been so blessed to have been a part of major international efforts to address carbon risks and identify green finance opportunities, I wanted to write this book to share my insights. I hope that you find it informative.

Acknowledgments

I will always be grateful to my parents, Dr. Jann Gumbiner and Dr. Albert Ezroj, who have supported me in all my endeavors.

My work has provided me with the great fortune of being able to collaborate with some of the world's trailblazers in addressing carbon risks and identifying green finance opportunities. To begin with, I would like to thank Dr. Evan Mills who theorized about the intersection of financial markets and climate change, collaborated on California's efforts at this intersection, and provided guidance on this book. Next, I would like to thank former California Insurance Commissioner Dave Jones, Geoffrey Summerhayes, and Butch Bacani, who have pioneered efforts to make the consideration of climate change a mainstream financial concern. Ideas that once may have seemed lofty are now a reality because of these individuals. Further, I would like to thank Ken Alex and Mary Nichols, who have led California climate change efforts, including California's delegation to the United Nations Climate Change Conference.

For help reviewing text, giving advice, and providing images for this book, I would like to thank the following modelers and policy experts: Natalie Ambrosio, Matthieu Bardout, Laura Biagioni, Marlena Chertock, Kajetan Czyz, Frank Elderson, Gerard Eljsink, Ethan Elkind, Nathan Fabian, Anton Gorodniuk, Vero Henze, Alice Hill, Henrik Jeppesen, Ted Lamm, Tianyi Luo, Leonardo Martinez-Diaz, Clare Murray, Jolien Noels, Betsy Otto, David Parham, Mike Peterson, David Pierce, Layalee Ramahi, Curtis Ravenel, Silvia Ruprecht-Martignoli, Leah Schleifer, Heffa Schücking, Aliona Tency, Jakob Thomä, Sylvain Vanston, Niek Verhoeven, and Susan Wilhelm. One of the things that I have enjoyed the most about my work is the many fascinating and dedicated individuals that I have gotten to meet. It is extremely flattering that people would take time out of their already busy schedules to further assist me with this book.

I would also like to thank the following individuals who have played a role in my pursuit of a career focused on public service and scholarship: Stephen Adams, Larry Agran, Fred Bercovitch, Susan Bernard, Anne Colten, Vincent Crawford, Marco Davanzo, Don Davis, Gary Downes, Janet Eichsteadt, Thomas Enslow, Stephen Gardbaum, Brian Hambler, Sean Hecht,

Bryant Henley, Hugh Hewitt, Amy Holm, Cara Horowitz, Gideon Kracov, Libio Latimer, Joel Laucher, Jack Lipton, Richard Mack, Jeremy McDaniels, Steve McEwen, Robert Nelson, Neil Peretz, Claudia Polsky, Arnie Rachlis, Valerie Ramey, Sandra Robbie, Bill Rubenstein, Kenny Russell, Valerie Sarfaty, Guy Scoffoni, Emily Smith, Nadia de Souza, Jules Stuyck, Will Swaim, Bruce Terry, Eric Vail, Geert Van Calster, Sam Wade, and Todd Walburg. I apologize to other individuals who have also helped me but whom I have not presently thanked. This book would not have been possible without the enormous support that I have received throughout my career.

Furthermore, I would like to thank my editorial team, including Kristina Abbotts, Christiana Mandizha, and Umamaheswari Chelladurai. I greatly appreciate their enthusiasm, professionalism, and skill.

1 Climate change becomes a financial concern

On a fall evening in 2015, one of the world's most esteemed bankers argued for addressing climate change. The head of the Bank of England, Governor Mark Carney, said the following at an exclusive business dinner gathering in London.

> There is a growing international consensus that climate change is unequivocal. Many of the changes in our world since the 1950s are without precedent: not merely over decades but over millennia.
>
> (Carney 2015)

In what started out as a speech one would expect from an environmentalist rather than a financial regulator, Governor Carney went on to describe how climate change presents serious financial risks. He explained that some carbon risks were already observable, such as losses from the increasing frequency and severity of weather events. Specifically, in just a few decades, weather-related loss events had increased by several times, and insurance losses had grown exponentially. This has led to insurers exiting certain markets and consumers having difficulty finding coverage (Carney 2015).

Still other carbon risks were lurking which could impact the global economy as a whole, including the possibility of a financial shock from a reevaluation of fossil fuel investments. According to Governor Carney, global monetary markets did not appear to be properly valuing and assessing the risks behind fossil fuel investments. Fossil fuel companies and those connected with them were being valued as if all available reserves could be exploited. However, if the world wanted to avoid catastrophic climate change, it was estimated that the majority of reserves could not be burned. If there was a sudden market correction acknowledging this, there could be a market shock, which some have referred to as the bursting of "carbon bubble." To avoid a tragic outcome, Governor Carney stated that the world's financial leaders needed to work together to ensure a smooth transition to a low-carbon economy (Carney 2015).

Later that year, with growing global concerns over the potential dire consequences of climate change, delegates from nearly two hundred countries met at an annual United Nations Climate Change Conference in Paris. At that meeting, delegates negotiated what is now referred to as the "Paris Agreement," where they committed to limit average global temperatures to no more than a 2 degrees Celsius increase above preindustrial levels and further sought a well below 2 degrees Celsius goal with a 1.5 degrees Celsius target (Paris Agreement 2015, Art. 2). Around this time, the United Nations also unveiled seventeen Sustainable Development Goals, with one of the major ones being climate action (UN 2015).

Following the Paris Agreement, which has been referred to as a tipping point in combating climate change (CDP n.d.), governments mobilized, including financial supervisors who make financial rules and regulators who enforce actions under these rules, starting to outline strategies for a smooth transition to a low-carbon economy. Governor Carney's Bank of England, the Bank of France, and the Dutch Bank all moved to formally recognize climate-related financial risks. Notable among the efforts of these pioneering financial regulators, the French government implemented the groundbreaking French Energy Transition Law. This law strengthened carbon disclosure requirements for companies operating in France and introduced them for institutional investors. In particular, the law required companies to disclose climate-related financial risks in annual reports, along with measures to reduce these risks. Moreover, the law required institutional investors to disclose in their annual reports how environmental considerations are taken into account when investing and how their policies are aligned with national transition strategies (French Energy Transition Law 2015, Art. 173).

The Chinese government pushed a powerful green agenda. This included developing the first government guidance for green loans and green bonds. Together these efforts quickly helped direct billions of dollars into sustainable green growth, transforming China into an international leader in green development (CBIRC 2012; PBOC 2015; NGFS 2019, 34).

California government entities identified thermal coal as an asset that could potentially become stranded and pushed for divestment (Jones 2016). The states' two largest pension funds divested from thermal coal. Then the state's insurance regulator asked insurers to voluntarily divest from thermal coal and started tracking thermal coal in insurers' investment portfolios.

With government agencies around the world starting to focus on a smooth transition to a low-carbon economy, collaborations among countries also started to emerge. With the insurance industry suffering some of the first climate-related financial losses, a global partnership emerged between insurance supervisors and regulators. Working with the United Nations, the United Kingdom, France, the Netherlands, and California partnered with other governments to launch the Sustainable Insurance Forum. Representing the first collection of financial supervisors and regulators of its kind, this group started convening meetings to discuss the sustainability challenges facing insurance supervisors and regulators and decided on best practices to address these challenges (SIF n.d.).

Meanwhile, the business community also took action. Companies expanded their participation in climate change efforts by becoming signatories of climate change initiatives, calculating and reporting greenhouse gas emissions, responding to surveys on carbon risks, divesting from or engaging regarding assets that could potentially become stranded, increasing their investments in green bonds, and working collaboratively with climate modelers. Previously niche efforts, these activities started becoming mainstream business activities.

Then, with both governments and businesses moving to recognize the need to ensure a smooth transition to a low-carbon economy, something pivotal happened. The two groups came together to develop overarching financial disclosure recommendations that could better inform the understanding of material risks. At the behest of Governor Carney and Michael Bloomberg, the G20 Financial Stability Board established the Task Force on Climate-related Financial Disclosures (TCFD), which included financial experts from Europe and the United States as well as other parts of the world (TCFD 2017a, 3).

The TCFD opened up a dialogue. Governments and companies that had already been active in this space commented on the TCFD recommendations as they were being prepared. Of note, the collection of insurance supervisors and regulators making up the Sustainable Insurance Forum filed comments jointly as financial supervisors and regulators. In comments titled "Leading Insurance Supervisors Support Adoption of Climate Disclosure Recommendations," the organization's supervisors and regulators explained the unique role that insurers play with disclosure and echoed the importance of the recommendations providing a comprehensive new standard for disclosing climate-related financial risks and opportunities (SIF 2017a).

In the middle of 2017, the TCFD released its final report. Now-TCFD Chair Bloomberg opened the report with a letter to Governor Carney stating that the recommendations will lead to "smarter, more efficient allocation of capital, and help smooth the transition to a more sustainable, low-carbon economy" (Bloomberg 2017, i). As explained in this letter, the recommendations will specifically help companies routinely consider the effects of climate change in business and investment decisions and disclose climate-related financial information. The letter closes with the statement:

> The risk climate change poses to businesses and financial markets is real and already present. It is more important than ever that businesses lead in understanding and responding to these risks – and seizing the opportunities – to build a stronger, more resilient, and sustainable global economy.
>
> (Bloomberg 2017, i)

The final report and supplemental documents themselves identify core elements along these lines, including governance, strategy, risk management, and metrics and targets. To begin with, on governance, the final report seeks the adoption of corporate governance strategies that account for carbon risks. In particular, the final report notes that it is especially important that the individuals making board and investment decisions be knowledgeable on climate risks and opportunities. Next, for strategy, the final report seeks transparency on the material financial risks and opportunities that have been identified over the short, medium, and long term. Further, with respect to risk management, the final report seeks information on the manner in which climate-related risks are managed and otherwise evaluated (TCFD 2017a, 13–23). Finally, with metrics and targets, the final report and supplemental documents detail different types of analytical tools that can inform corporate governance and facilitate disclosure. Principal among these tools is discussion of scenario analysis. This tool allows for projecting the impacts of climate change based on the modeling of certain possible scenarios. According to the report, a key means for using this tool is to determine alignment with the objectives of the Paris Agreement. On the one hand, such a tool can evaluate the impact on financial markets as economies move toward Celsius targets of 2 degrees or below. On the other hand, such a tool can also be used to evaluate the impact of an increased frequency and severity of climate-related weather disasters if economies fail to meet these targets (TCFD 2017a, 25–30, 2017b). Figure 1.1 explains the core elements of the TCFD recommendations.

Following publication of the TCFD final report, governments and businesses began centering their messaging around the TCFD recommendations. The collection of international insurance supervisors and regulators making up the Sustainable Insurance Forum released a Sustainable Insurance Forum statement in support of the TCFD recommendations, saying that

Figure 1.1 Core Elements of Recommended Climate-related Financial Disclosures

Source: Task Force on Climate-related Financial Disclosures

they welcomed the recommendations and looked forward to working to make them a reality (SIF 2017b). Further, the supervisors and regulators specifically identified aspects of corporate governance, such as the application of the scenario analysis tool, as main components of their intended efforts to support the TCFD recommendations (SIF n.d.). The Sustainable Insurance Forum also later published a joint report with the international insurance standards setting organization, the International Association of Insurance Supervisors, on how the insurance sector has been incorporating the TCFD recommendations (IAIS and SIF 2018).

French President Emmanuel Macron convened a One Planet Summit with global climate leaders. Here international banking supervisors and regulators established the Central Banks and Supervisors Network for Greening the Financial System (NGFS 2017). The NGFS's stated purpose has been to strengthen the global response that is necessary to meet the Paris Agreement objectives and to increase the financial system's role in managing risks and mobilizing capital for green and low-carbon investments. The group has set up three dedicated workstreams: supervision, macrofinancial, and mainstreaming green finance, which involves mobilizing capital for environmentally sustainable activities. The NGFS later published a report encouraging corporate disclosure in line with the TCFD recommendations (NGFS 2019).

Around the same time, California issued its *Trial by Fire* report highlighting the impact that climate change is having on wildfires and the risk that climate change presents to over five trillion in actively managed assets held by insurance companies operating in California (Mills et al. 2018). In the foreword, then-California Insurance Commissioner Dave Jones explained:

> Disclosure of climate-related risks to the financial sector is important for the sustainability of the global and United States financial systems generally, and the insurance sector in particular. In 2017, the G-20 Financial Stability Board's Task Force on Climate-related Financial Disclosures issued its recommendations for climate risk disclosures for each economic sector, including the insurance sector. Insurers, investors, policyholders, and regulators should all support and implement these recommendations.
>
> (Jones 2018, ii)

The California Department of Insurance and the Swiss Ministry of Environment also published initial widescale applications of scenario analysis on assets in investment portfolios. Findings were featured in an article in *The Economist* with a self-explanatory title: "Markets may be underpricing climate-related risk" (Piotrowski 2018).

A growing number of companies also publicly endorsed the TCFD recommendations by signing up as supporters and publishing reports that either directly or indirectly referenced the recommendations. For instance, companies released reports explaining how they addressed climate change and labeled them TCFD reports or TCFD and French Energy Transition Law reports when doing business in France. Companies examined greenhouse gas emissions through methodologies outlined in the TCFD recommendations. Companies continued or started to provide information through surveys that had been updated to track the recommendations. Companies also performed scenario analysis on their investment portfolios. A status update published in 2018 announced that over five hundred companies and other organizations, including numerous Fortune 500 companies, had become official supporters of the recommendations (TCFD 2018, iv).

With the financial community embracing a unifying set of recommendations, there was a lot of optimism. However, this optimism was checked at the end of 2018 with United Nations Intergovernmental Panel on Climate Change (IPCC) and United States National Climate Assessment reports providing bleak projections for the impact of climate change on the world (USGCRP 2018; IPCC 2018). For instance, according to the IPCC, a 2 degrees Celsius increase would result in irreparable global damage, such as the increased severity of weather-related catastrophes, significant sea level rise, and the destruction of sensitive ecosystems. Limiting global warming to 1.5 degrees Celsius would stem some of this damage but would only be possible with "large, immediate and unprecedented global efforts" to mitigate greenhouse gas emissions (IPCC 2018, 36). With the publication of these reports, it became generally accepted that meeting a 2 degrees Celsius target was insufficient. Efforts refocused instead on achieving a below 2 degrees Celsius target, specifically a 1.5 degrees Celsius target (CDI and 2DII 2019; AXA 2019; Climate Risk Disclosure Act 2019).

Nevertheless, major economies around the world were undeterred in their commitment to addressing carbon risk and have now also been pushing for important green finance efforts. This has continued even with the global COVID-19 pandemic. The United States House of Representatives is discussing having the United States Securities and Exchange Commission mandate carbon risk disclosures while also entertaining plans for an expansive Green New Deal. The governing body of the European Union, the European Commission, began implementing its comprehensive strategy to address carbon risk and identify green finance opportunities through its action plan on sustainable finance (EC 2018). The China Securities Regulatory Commission is developing requirements for carbon risk disclosure, while the China Banking and Insurance Regulatory Commission has been pushing standards for green loans, and the People's Bank of China has been pushing a Green Bond Endorsed Project Catalogue.

Meanwhile, analytical tools have continued to advance. Reporting frameworks are harmonizing to allow for TCFD disclosures. The European Commission is managing a green taxonomy to classify environmentally sustainable activities and reduce barriers to investing in those activities. California expanded its scenario analysis to look at below 2 degrees Celsius targets and to include a physical risk assessment. The Bank of England published extreme weather event simulations that can be used in catastrophe modeling to evaluate underwriting (Bank of England 2019).

The success of major economies in achieving their larger ambitions will depend on the application of analytical tools. First, reporting frameworks can allow companies to see how well they are addressing carbon risks, in particular with respect to the TCFD recommendations. Second, by quantifying how much greenhouse gas companies emit into the atmosphere as a direct or indirect result of their operations, carbon footprint calculations can help identify carbon risks with particular companies, especially within supply chains. Third, brown taxonomies can help investors identify current carbon risks by classifying fossil fuel assets in a systematic manner. Fourth, green taxonomies can help investors identify current green finance opportunities by classifying sustainable activities in a systematic manner. Fifth, scenario analysis for assets can help investors identify future carbon risks and green finance opportunities. Finally, stress testing for liabilities can help insurers and banks address future carbon risks and better inform policymakers.

References

AXA. 2019. *2019 Climate Report*. Paris: AXA.

Bank of England. 2019. *General Insurance Stress Test 2019: Scenario Specification, Guidelines and Instructions*. London: Bank of England.

Bloomberg, Michael. 2017. *Foreword to Recommendations of the Task Force on Climate-related Financial Disclosures*. Edited by Task Force on Climate-related Financial Disclosures, i. Basel: Task Force on Climate-related Financial Disclosures.

California Department of Insurance and 2 Degrees Investing Initiative. 2019. *2 Degrees Scenario Analysis, Insurance Companies Operating in California*. Sacramento: California Department of Insurance and 2 Degrees Investing Initiative.

Carney, Mark. 2015. "Breaking the Tragedy of the Horizon – Climate Change and Financial Stability." Speech at Lloyd's of London, London, September.

CDP. n.d. "CDP Climate Change Questionnaire." Accessed July 25, 2020. https://guidance.cdp.net/en/guidance?cid=2&ctype=theme&idtype=ThemeID&incchild=1%C2%B5site%3D0&otype=Questionnaire&page=3&tags=TAG-646%2CTAG-605%2CTAG-600.

China Banking and Insurance Regulatory Commission. 2012. *Guidance on Green Loans*. Beijing: China Banking and Insurance Regulatory Commission.

Climate Risk Disclosure Act of 2019. H.R.3623 (2019).

European Commission. 2018. *Commission Action Plan on Financing Sustainable Growth*. Brussels: European Commission.

French Energy Transition Law, 2015–992 (2015).

Intergovernmental Panel on Climate Change. 2018. *Global Warming of 1.5°C*. Geneva: Intergovernmental Panel on Climate Change.

International Association of Insurance Supervisors and Sustainable Insurance Forum. 2018. *Issues Paper on Climate Change Risks to the Insurance Sector*. Geneva: International Association of Insurance Supervisors and Sustainable Insurance Forum.

Jones, Dave. 2016. *Coal Divestment & Carbon-Based Investment Data Call*. Sacramento: California Department of Insurance.

Jones, Dave. 2018. *Foreword to Trial by Fire: Managing Climate Risks Facing Insurance in the Golden State*. Edited by Evan Mills, Ted Lamm, Sadaf Sukhia, Ethan Elkind, and Aaron Ezroj, ii. Sacramento: California Department of Insurance.

Mills, Evan, Ted Lamm, Sadaf Sukhia, Ethan Elkind, and Aaron Ezroj. 2018. *Trial by Fire: Managing Climate Risks Facing Insurance in the Golden State*. Sacramento: California Department of Insurance.

Network for Greening the Financial System. 2017. *Joint Statement by the Founding Members of the Central Banks and Supervisors Network for Greening the Financial System*. Paris: Network for Greening the Financial System.

Network for Greening the Financial System. 2019. *A Call for Action, Climate Change as a Source of Financial Risk*. Paris: Network for Greening the Financial System.

Paris Agreement to the United Nations Framework Convention on Climate Change, December 12, 2015, T.I.A.S. No. 16–1104.

People's Bank of China's Green Finance Committee of the China Society for Finance and Banking. 2015. *Green Bond Endorsed Project Catalogue*. Beijing: People's Bank of China's Green Finance Committee of the China Society for Finance and Banking.

Piotrowski, Jan. 2018. "Markets May be Underpricing Climate-related Risk." *The Economist*, May 24. www.economist.com/finance-and-economics/2018/05/24/markets-may-be-underpricing-climate-related-risk.

Sustainable Insurance Forum. 2017a. *Leading Insurance Supervisors Support Adoption of Climate Disclosure Recommendations*. Geneva: Sustainable Insurance Forum. *Comments*, February.

Sustainable Insurance Forum. 2017b. *Leading Insurance Supervisors Support Adoption of Climate Disclosure Recommendations*. Geneva: Sustainable Insurance Forum. *Statement*, July.

Sustainable Insurance Forum. n.d. "Sustainable Insurance Forum." Accessed July 25, 2020. www.sustainableinsuranceforum.org.

Task Force on Climate-related Financial Disclosures. 2017a. *Recommendations of the Task Force on Climate-related Financial Disclosures*. Basel: Task Force on Climate-related Financial Disclosures.

Task Force on Climate-related Financial Disclosures. 2017b. *The Use of Scenario Analysis in Disclosure of Climate-Related Risks and Opportunities*. Basel: Task Force on Climate-related Financial Disclosures.

Task Force on Climate-related Financial Disclosures. 2018. *Task Force on Climate-related Financial Disclosures: Status Report*. Basel: Task Force on Climate-related Financial Disclosures.

United Nations. 2015. *2030 Agenda for Sustainable Development*. New York: United Nations.

United States Global Change Research Program. 2018. *Impacts, Risks, and Adaptation in the United States: Fourth National Climate Assessment, Volume II*. Washington, DC: United States Global Change Research Program.

2 Reporting frameworks

Reporting frameworks can allow companies to see how well they are addressing carbon risks, in particular with respect to the TCFD recommendations.

The global financial community consists of a number of public and private actors. To begin with, there is the Financial Stability Board that includes the G20 major economies and ultimately endorsed the development of the TCFD. There are also more specialized international banking, insurance, and investor international government collaborations. Within countries, there are financial supervisors and regulators either at the national or subnational level. Among private actors, there are banks that lend capital to consumers for activities such as purchasing private property. There are insurers that issue policies to consumers and reinsurers that hold portions of insurers' risks for paying out on those policies. There are rating agencies that evaluate these companies to assess their creditworthiness. There are also modelers and third-party data providers that develop and sell analytical tools to these companies.

When the global financial community turned its attention to evaluating carbon risks, analytical tools were already being used. However, those tools were generally not tailored for financial institutions. While informative, the tools allowed for significant variation and predominantly qualitative assessments. This was dissimilar from more precise and quantitative tools typically relied on within the financial community. More recently, though, analytical tools, such as reporting frameworks, have become increasingly sophisticated, with comparable metrics across users.

Reporting frameworks have always had value because, at a minimum, they encouraged companies to internally convene, collect information, set targets, and track progress. However, with more comparable metrics, reporting frameworks are allowing companies to see how they are performing with respect to their peers and garner information on best practices with a range of analytical tools. Reporting frameworks now ask for advanced carbon footprint metrics, allow for discussion on divestment and engagement using brown taxonomies and investment using green taxonomies, and seek information on scenario analysis for assets and stress testing for liabilities.

2.1 Development

With companies starting to track their carbon footprints, a nonprofit organization, CDP (formerly the Carbon Disclosure Project), realized that there was value in assembling this information in a central location. It started collecting carbon footprints from companies, building the world's largest database of carbon footprints. Over time, realizing that other climate-related information was being assembled and had practical value, CDP increasingly sought additional climate-related information from companies. In the process, CDP developed a voluntary reporting framework for companies wishing to report on risks and opportunities presented by climate change.

While the CDP methodology has become somewhat of an industry standard, with other reporting frameworks even emulating it, today there are a number of reporting frameworks covering various sectors and having different focuses. There are a number of established voluntary frameworks managed by nonprofit organizations, such as CDP, the Global Reporting Initiative (GRI), and the Climate Disclosure Standards Board (CDSB) with the Sustainability Accounting Standards Board (SASB). There are also mandatory frameworks that government financial regulators have been developing, in particular a framework developed for large companies operating in France and insurance companies operating in certain jurisdictions within the United States.

2.2 Harmonization with Task Force on Climate-related Financial Disclosures recommendations

Major reporting frameworks are now embracing the TCFD recommendations. Although reporting frameworks began with a focus on collecting carbon footprints, and the collection of this data remains important, reporting frameworks now seek various metrics on how companies are responding to climate change. The questions are being mapped to the TCFD recommendations. In fact, several of the leading reporting framework developers, including CDP, GRI, and CDSB with SASB, are even collaborating to further map their reporting frameworks to the TCFD recommendations, which cover governance, strategy, risk management, and metrics and targets, as detailed in Table 2.1.

2.2.1 CDP climate change questionnaire

When CDP started collecting climate information in 2002, there were a couple hundred companies providing information. However, now, thousands of companies are reporting through CDP (CDP n.d.a). This accounts for nearly sixty percent of global market capitalization and a significant portion of the world's greenhouse gas emissions (GRI and CDP 2017).

Table 2.1 TCFD Recommended Disclosures

Governance	*Strategy*	*Risk Management*	*Metrics and Targets*
Disclose the company's governance around climate-related risks and opportunities. a) Describe the board's oversight and b) Describe management's role.	Disclose the actual and potential impacts of climate-related risks and opportunities on the company's business, strategy, and financial planning where such information is material. a) Describe risks and opportunities that have been identified over the short, medium, and long term and b) Describe the company's strategy when considering climate scenarios such as 2 degrees or below 2 degrees Celsius scenarios.	Disclose how the company identifies, assesses, and manages climate-related risks. a) Describe the company's processes for identifying, assessing, and managing risks and b) Describe how these processes are integrated into the company's overall risk management.	Disclose the metrics and targets used to assess and manage relevant climate-related risks and opportunities where such information is material. a) Disclose metrics used by the company and b) Describe the targets used by the company (TCFD 2017, 14).

Following a redesign in 2018, the online CDP climate change questionnaire provides a good example of what kind of information is sought by reporting frameworks and how a reporting framework can map to the TCFD recommendations (CDP n.d.b). First, the questionnaire itself both begins and ends with corporate governance questions. Questions get at who exactly within a company is making decisions on climate-related issues and what their authority is within the company, and whether there are rewards and targets related to any climate-related issues. The questionnaire also asks for details on the person who has approved responses, specifically asking for a job title and the selection of a job category that ranges in status. For some of these questions, a link is specifically drawn to the TCFD recommendations by identifying relevant TCFD recommendations on corporate governance, such as recommended disclosures on the board's oversight of climate-related risks and opportunities and management's role in assessing and managing climate-related risks and opportunities (CDP n.d.b).

Second, a focus of the questionnaire is how companies are identifying risks and opportunities, specifically with respect to a 2 degrees Celsius scenario. Questions in this area allow a company to describe how it identifies short-, medium-, and long-term climate-related issues. A company is asked to discuss how climate-related issues are integrated into its business strategy, specifically asking about the use of scenario analysis and what scenarios and other inputs are being used for this analysis. For some of these questions, a link is specifically drawn to the TCFD recommendations by identifying relevant TCFD recommendations on strategy, including describing the climate-related risks and opportunities that a company has identified over the short, medium, and long term; describing the impact of climate-related risks and opportunities on the company's business, strategy, and financial planning; and describing the resilience of the company's strategy, taking into consideration different climate scenarios, including a 2 degrees Celsius or lower scenario (CDP n.d.b).

Third, there are questions on how a company is responding to specific risks (e.g., current and emerging regulation, technological advancements, legal concerns, market shifts, reputational considerations, acute and chronic physical stress, and upstream and downstream challenges) and specific opportunities (e.g., increasing resource efficiency, identifying new energy sources, creating additional products and services, market shifts, and improving resilience). The questionnaire further probes whether a company has been employing specific desirable risk management strategies. For example, the questionnaire seeks information on energy consumption practices, value chain assessments, and publication of climate change information in places other than its CDP response. The questionnaire also has optional questions on supply chain assessments for products and services (CDP n.d.b).

Fourth, the questionnaire asks how a company is using metrics and targets, with an emphasis on carbon footprints. CDP maintains the largest collection of companies' carbon footprints. Therefore, it is not surprising that large portions of its questionnaire are dedicated to collecting and evaluating how a company tracks greenhouse gas emissions. These areas of the questionnaire allow a company to provide Scope 1 emissions, which are direct emissions from owned or controlled sources; Scope 2 emissions, which are indirect emissions from the purchase of electricity; and, where available, Scope 3 emissions, which are indirect emissions that occur in the value chain. The questionnaire asks for a breakdown of Scope 1 and Scope 2 emissions by greenhouse gas type for the six greenhouse gases that are restricted by the international Kyoto Protocol treaty: carbon dioxide, methane, nitrous oxide, hydrofluorocarbons, perfluorocarbons, and sulfur hexafluoride. The questionnaire asks for a breakdown of Scope 1 and Scope 2 emissions by country/region, business division, facility, and activity. Moreover, the questionnaire asks for carbon intensity metrics, which are advanced carbon footprint metrics that quantify emissions in relation to business and investment operations. The questionnaire also asks a company to identify standards, protocols, and methodologies used to calculate emissions, such as the Greenhouse Gas Protocol: A Corporate Accounting and Reporting Standard (CDP n.d.b). The questionnaire asks about year over year comparisons of Scope 1 and Scope 2 emissions and reasons for any changes. The questionnaire further asks for verification for emissions data and to discuss any emissions targets and progress against those targets (CDP n.d.b). For some carbon footprint questions, a link is specifically drawn to the TCFD recommendations by identifying relevant TCFD recommendations on metrics and targets, including disclosing the metrics used by the company to assess climate-related risks and opportunities in line with its strategy and risks management process, disclosing Scope 1, Scope 2, and, if appropriate, Scope 3 greenhouse gas emissions and related risks and describing the targets used by the company to manage climate-related risks and opportunities and performance against targets (CDP n.d.b). More than ninety percent of the Fortune 500 companies reporting to CDP used the GHG Protocol Corporate Standard or a methodology based on the GHG Protocol Corporate Standard to report their emissions (GHG Protocol 2020).

As far as structure, the CDP climate change questionnaire allows for a broad range of companies to provide responses. However, it can also be tailored to specific types of companies, taking into account any sectors that the company operates in and their ability to provide comprehensive responses. When first entering information for a company, the CDP website prompts whoever is entering data to identify whether the company operates in one or more of a dozen high-impact sectors, which will prompt the user to answer additional questions for those sectors. The CDP website then allows the company to select a minimum or full version. The minimum version contains identical but fewer questions and no sector-specific questions. CDP states that the minimum version can be used by smaller companies or by companies responding to the questionnaire for the first time. A company can also decide to respond to additional supply chain questions. The survey has yes/no, multiple-choice, and open-text questions, allowing for comprehensive, comparable data (CDP n.d.b).

After completing the CDP climate change questionnaire, the person who has entered information for a company will receive an email with scored results for the company. Not all information is made publicly available, but public responses for thousands of companies can still be viewed on the CDP website. CDP also lists over one hundred companies identified as "climate change A list," an identification which is good for a company's reputation (CDP n.d.b).

2.2.2 *Global Reporting Initiative Sustainability Reporting Standards*

Prior to the rollout of the CDP questionnaire, nonprofit organizations, including the Ceres and the United Nations, formed GRI. The GRI has developed Sustainability Reporting Standards (formerly referred to as "Guidelines"). These were first published in 2000 and now have thousands of users that rely on them to publish sustainability reports (GRI 2020).

In 2016, the GRI Sustainability Reporting Standards were redone. This redrafting more closely aligns them with the TCFD recommendations that were being developed at the time. Major changes have had to do with an increased emphasis on management's approach to material disclosures and consideration of supply chain impacts. More minor changes related to modifying terminology so that it more closely reflects usage in the international financial community, such as using the term "disclosure" rather than "indicator" (GRI 2020).

The redrafting of the GRI Sustainability Reporting Standards also made them easier to use. The GRI Sustainability Reporting Standards now have a modular structure (GRI 2020), making it easier for a company to update information, since attention may only be needed in certain areas, and allowing for reporting only on specific areas. Moreover, the modular structure also allows for reporting on specific areas rather than a single comprehensive report (GRI 2020). Figure 2.1 shows the current modular structure of the GRI Sustainability Reporting Standards.

As shown in the figure, the modular structure includes three universal standards and a number of topic-specific standards. A comprehensive report would include a report with information for the three universal standards and information on applicable areas of topic-specific standards. A report on a specific area would instead include a report on an applicable topic-specific

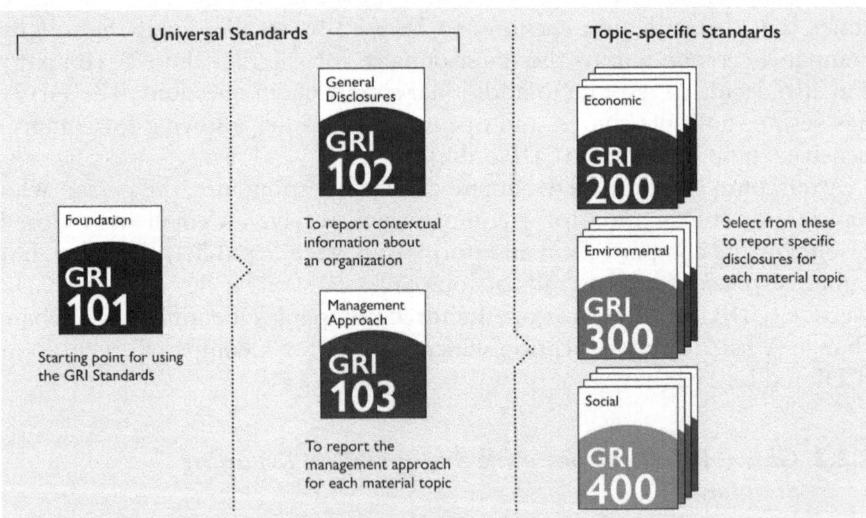

Figure 2.1 Modular Structure

Source: Global Reporting Initiative. This image was viewable on July 25, 2020 on the following GRI website: www.globalreporting.org/information/news-and-press-center/Pages/Ready-for-GRI-Standards-The-release-date-is-announced!-.aspx. Per GRI, GRI is the independent international organization – headquartered in Amsterdam with regional offices around the world – that helps businesses, governments, and other organizations understand and communicate their sustainability impacts.

standard with management approach information discussing material information related to that specific standard (GRI 2020).

The GRI Sustainability Reporting Standards have a topic-specific standard for greenhouse gas emissions. The standard requires separately reporting gross Scope 1, Scope 2, and Scope 3 emissions. In doing so, the standard seeks information on the following greenhouse gases: carbon dioxide, methane, nitrous oxide, hydrofluorocarbons, perfluorocarbons, sulfur hexafluoride, and nitrogen trifluoride (GRI and GSSB 2016e, 4). The standard also seeks carbon intensity information (GRI and GSSB 2016e, 13). Moreover, the standard seeks information on methodologies, assumptions, and calculation tools that were used in composing greenhouse information (GRI and GSSB 2016e, 7–8, 11–13). The standard further asks for information on how the company's initiatives have reduced greenhouse gas emissions (GRI and GSSB 2016e, 14).

Greenhouse gas emissions reporting through the GRI is similar in practice to reporting through CDP. CDP and the GRI worked closely together to ensure harmonization of greenhouse gas emission reporting, with CDP being an active contributor to the revision of GRI's greenhouse has gas emissions disclosures. Harmonization has been achieved because both reporting frameworks rely extensively on the GHG Protocol Corporate Standard and GHG Protocol Scope 3 Corporate Standard (GRI and CDP 2017, 5). There are some distinctions between the two reporting frameworks, though. In particular, the GRI requires reporting Scope 3 emissions and an additional greenhouse gas beyond those restricted in the international Kyoto Protocol treaty, where doing so is voluntary with the CDP climate change questionnaire, although doing so would presumably lead to more favorable assessments of the reporting practices of companies in the CDP climate change questionnaire.

Other topic-specific standards are relevant to reporting on carbon risks and green finance opportunities. An energy standard seeks information on total energy consumption (GRI and GSSB 2016d, 6–7). An economic performance standard specifically seeks information on the financial implications of corporate actions (GRI and GSSB 2016c, 9–10). A public policy standard asks for information on public policy outreach, covering areas such as the management of carbon risks and potentially green finance (GRI and GSSB 2016f, 5–6).

The management approach standard is extremely relevant to reporting on carbon risks and potentially green finance. This standard asks for information on material topics, which GRI defines as topics that reflect "significant economic, environmental and social impacts, or that substantively influences the assessments and decisions of stakeholders" (GRI and GSSB 2016b, 12). For each material topic, this standard seeks information on how a company manages the topic, including information on specific actions and goals and targets (GRI and GSSB 2016b, 6–11).

The general disclosures standards are also relevant to reporting on carbon risks and potentially green finance when preparing a comprehensive report.

This standard seeks a statement from the company's most senior decision maker, such as the company's CEO, on the relevance of sustainability to the companies and the company's sustainability strategy (GRI and GSSB 2016a, 14). This standard also seeks information on the process for delegating authority on sustainability issues (GRI and GSSB 2016a, 18). Further, this standard seeks information on the general composition of companies, including supply chain information (GRI and GSSB 2016a, 9–12).

The GRI allows companies to publish a sustainability reports that cover their entire Sustainability Reporting Standards or a limited scope of their Sustainability Reporting Standards. Companies that publish a sustainability report that covers a limited scope of the GRI Sustainability Reporting Standards are asked to identify that they are doing such. All companies using the GRI Sustainability Reporting Standards to publish sustainable reports are asked to report doing so to the GRI so that the GRI can track this, even though there are no costs associated with using the GRI Sustainability Reporting Standards (GRI 2020).

2.2.3 Climate Disclosure Standards Board framework using Sustainability Accounting Standards Board standards

SASB is an organization that has developed investor-focused reporting standards for financially material environmental disclosures. In 2018, working with CDSB, SASB developed implementation guidance describing how SASB standards and the CDSB reporting framework can be used together to make TCFD disclosures. Doing so allows sections of the reporting framework, referred to as core elements, to specifically align with the TCFD recommendations. For instance, the reporting framework includes sections on governance, strategy, risk management, and metrics and targets (CDSB and SASB 2019, 17–55). SASB states that a company using its standards with the CDSB reporting framework is well positioned to make TCFD disclosures. A SASB and CDSB implementation guide explains that the TCFD recommendations sought to establish common principles to help existing disclosure regimes more closely align over time. According to this guide,

> [b]y facilitating more streamlined, more effective disclosure of financially material, climate-related information, the TCFD recommendations – implemented using the SASB standards and CDSB [f]ramework – can help companies better meet the changing needs of investors, lenders, and insurance underwriters, while also supporting the efficient functioning of the markets, as well as the long-term stability and resilience of the global economy.
>
> (CDSB and SASB 2019, 60)

The design of the CDSB reporting framework using SASB standards resembles that of other prominent climate change reporting frameworks.

Similar to how the GRI and CDP reporting frameworks operate, a company has the ability to provide both qualitative and quantitative responses. However, this reporting framework is tailored to specific industries and focused on financial material disclosures (CDSB and SASB 2019, 17–55).

2.2.4 *French Energy Transition Law*

In addition to nonprofit climate change reporting frameworks, there are also government reporting frameworks that require mandatory disclosures. For example, in 2015, the French government adopted the world's first national climate disclosure law, the French Energy Transition Law. The law went into effect the following year. The law requires companies to consider climate-related financial risks and report on how they are doing so (French Energy Transition Law 2015, Art. 173; 2DII 2015, 7; Mason et al. 2016).

The French Energy Transition Law does not currently map to the TCFD recommendations, but it covers some of the main recommendations. Therefore, a company can satisfy its requirements under the law through TCFD-aligned reporting. To begin with, with respect to strategy, the law asks a company for information on how climate change will present it with financial risks. Next, with respect to risk management, the law asks a company about any low-carbon strategies it is adopting to reduce these risks, in particular with the production of goods and provision of services. The law further requires investors to make certain financial information available to their beneficiaries. This includes information on how their investment decisions incorporate climate-related risks such as greenhouse gas emissions associated with their assets and information on actions taken to contribute to a transition to a low-carbon economy. Furthermore, with respect to metrics and targets, the law requires banks and credit institutions to disclose risks of excessive leverage that are shown through regularly implemented stress tests. If appropriate, institutional investors shall explain why contributions did not meet targets from a prior fiscal year. As far as structure, the French government has stated that listed companies shall disclose this qualitative information in an annual report (French Energy Transition Law 2015, Art. 173; 2DII 2015, 7; Mason et al. 2016). Collected information fits into the core elements of the TCFD-recommended climate-related financial disclosures as follows.

2.2.5 *National Association of Insurance Commissioners Climate Risk Disclosure Survey*

Some United States government agencies at the state level also have a sector-specific reporting framework, specifically those regulating the insurance sector. Several states in the United States have been requiring larger insurers operating in their jurisdiction to respond to a survey that has been organized in connection with the National Association of Insurance Commissioners

(NAIC), the NAIC Climate Risk Disclosure Survey, modeled at least in part on an earlier version of the CDP climate change questionnaire (NAIC n.d.). The survey and supplemental guidance currently do not specifically map to the TCFD recommendations but cover many areas of the TCFD recommendations, and the most recent survey was sent to insurers with a cover letter explaining how the questions correspond to the TCFD recommendations (IAIS and SIF 2020, 37). To begin with, on governance, the survey is sent directly to the CEOs of insurers, and supplemental guidance asks about the role of the board of directors in managing carbon risk (CDI 2018, 2). Next, with respect to strategy, the survey seeks information on the current or anticipated risks that climate change presents to a company and the degree that climate change risks could impact a company's business, including financial implications (CDI 2018, 3). Further, with respect to risk management, the survey asks about efforts to mitigate greenhouse gas emissions, investment strategies, preventing losses from climate change-influenced events, and public engagement (CDI 2018, 2–5). Finally, with respect to metrics and targets, the survey asks that a company describe actions being taken to manage the risks climate change poses to businesses, including, in general terms, the use of computer modeling (CDI 2018, 5). Supplemental guidance discusses measuring carbon footprints using the GHG Protocol (CDI 2018, 2). Collected information fits into the core elements of the TCFD recommended climate-related financial disclosures as follows.

As far as structure, the NAIC Climate Risk Disclosure Survey allows a company to provide qualitative responses to eight questions covering areas such as efforts to mitigate greenhouse gas emissions, investment strategies, preventing losses from climate change-influenced events, public engagement, and modeling (CDI n.d.). California, which is now taking the lead in implementing the survey, posts the survey results on a searchable database. A nonprofit investor organization, Ceres, has used the publicly available survey information to identify best actors (Messervy and McHale 2016). Results were also cited in California's *Trial by Fire* report (Mills et al. 2018, 25–27, 32, 64–65, 67, 70, 73) and the joint TCFD paper of the Sustainable Insurance Forum and International Association of Insurance Supervisors (IAIS and SIF 2018).

2.2.6 *Comparing reporting frameworks*

When comparing the more established climate change reporting frameworks, there are observable trends. To begin with, nearly all the reporting frameworks ask for information on board oversight. Next, all the reporting frameworks ask how carbon risks will impact a company, and almost all ask about opportunities, too. Nevertheless, not all reporting frameworks ask that the information be divided into disclosures on short-, medium-, and long-term impacts. Further, all the reporting frameworks seek information on how companies are managing carbon risks. Moreover, all the reporting frameworks also ask about metrics and targets, at least in certain instances. The metrics that are specifically identified include carbon footprints, scenario analysis, and stress tests. Table 2.2 compares the content of these different reporting frameworks.

However, while the reporting frameworks seek a lot of similar information, there are some important structural distinctions. First, some reporting frameworks require a submission to the entities overseeing the implementation of the framework. However, others encourage a company itself to publish a sustainability report or direct a company to provide disclosures in an annual report. Second, the reporting frameworks are for different types of companies. Most of the voluntary reporting frameworks are set up for all companies. The mandatory reporting frameworks, however, are for specific types of business. This is because the entity administering the framework has chosen a subset of businesses either to facilitate compliance or to ensure that the entity administering the framework has proper authority. Table 2.3 provides a comparison of the structure of some of the major climate change reporting frameworks.

Building on information sought by and provided through reporting frameworks, London School of Economics researchers have made a major breakthrough in tracking the efforts of companies to address carbon risks. These researchers developed a Transition Pathway Initiative tool that scores a company's efforts to address carbon risks on a staircase using publicly available information gathered by a third-party data provider. The Transition Pathway Initiative tool assigns climate action ranks based on a company's efforts, ranging from Levels 0 to 4. Various indicators are used to determine where a company fits within the ranges from Levels 0 to 4. As with the developers of reporting frameworks, developers of the Transition Pathway Initiative tool have mapped these indicators against the TCFD

Table 2.2 Reporting Framework Content Comparison

	Governance	*Strategy*	*Risk Management*	*Metrics and Targets*
CDP Climate Change Questionnaire	Yes	Yes	Yes	Yes
GRI Sustainability Reporting Standards	Yes	Yes	Yes	Yes
CDSB Framework Using SASB Standards	Yes	Yes, without timelines	Yes	Yes
French Energy Transition Law	No	Yes, without timelines	Yes	Yes, in certain instances
NAIC Climate Risk Disclosure Survey	Yes	Yes, for risks without timelines	Yes	Yes

Table 2.3 Reporting Framework Structural Comparison

	Information	*Method*	*Applicability*	*Requirement*
CDP Climate Change Questionnaire	Qualitative and Quantitative	Submit to CDP	All Businesses	Must Submit to Receive a Rating
GRI Sustainability Reporting Standards	Qualitative and Quantitative	Publish a Sustainability Report	All Businesses	Voluntary
CDSB Framework Using SASB Standards	Qualitative and Quantitative	Provide in Annual Report	All Businesses	Voluntary
French Energy Transition Law	Qualitative	Provide in Annual Report	Listed Businesses in France	Mandatory
NAIC Climate Risk Disclosure Survey	Qualitative	Submit to Regulator	Insurers with US$100 Million or More in Annual Premium Licensed in California, Connecticut, Minnesota, New Mexico, New York, and Washington	Mandatory

recommendations (Dietz, Jahn et al. 2019, 4–13). This ensures that rankings appropriately consider governance, strategy, risk management, and metrics and targets.

To begin with, a company is ranked at Level 0 when it is unaware of or not acknowledging climate change as a business issue. In particular, a company is ranked at Level 0 when it is not taking one of several actions that show that it is acknowledging climate change as a significant issue for business. This includes, for instance, adopting a climate change policy or making a statement committing it to take action on climate change (Dietz, Jahn et al. 2019, 9).

Next, a company is ranked at Level 1 when it is acknowledging climate change as a business issue but is not necessarily taking tangible steps to address climate change. In particular, a company is ranked at Level 1 when it does both of the following. First, the company adopted a climate change policy or has made a statement committing it to take action on climate change. Second, the company incorporated at least two of several advanced management practices demonstrating that it is recognizing climate change as a relevant risk or opportunity to its business (Dietz, Jahn et al. 2019, 10).

Further, a company is ranked at Level 2 when it develops basic capabilities for addressing climate change, including management strategies and reporting procedures. A company is ranked at Level 2 when it does everything required at Level 1 and does both of the following. First, the company has set greenhouse gas emission reduction targets. Second, the company has published information on its Scope 1 and Scope 2 greenhouse gas emissions (Dietz, Jahn et al. 2019, 10–11).

Moreover, a company is ranked at Level 3 when it is integrating addressing climate change into operation decision making. A company is ranked at Level 3 when it does everything required at Levels 1 and 2 and takes several significant actions to address climate change. Among other significant actions, this includes assigning senior management or board responsibilities for climate change and providing comprehensive disclosures that include verified Scope 1, 2, and 3 emissions and information on climate change policy efforts (Dietz, Jahn et al. 2019, 11–12).

Finally, a company is ranked at Level 4 when it develops a more strategic and comprehensive understanding of the risks and opportunities related to a low-carbon transition and incorporates this into its business strategy and capital expenditure decisions. In particular, a company is ranked at Level 4 when it does everything required in Levels 1 through 3 and takes several significant calibrated actions to address climate change. The several significant calibrated actions to address climate change include setting long-term quantitative targets for reducing greenhouse gas emissions, compensating senior executives who incorporate climate change performance into decision making, using scenario analysis, and making sure that trade associations that the company participates in do not take positions that are inconsistent with the company's climate change policy (Dietz, Jahn et al. 2019, 12).

The Transition Pathway Initiative tool ranks a company as a "four star" company when it achieves all the Transition Pathway Initiative indicators (Dietz, Jahn et al. 2019, 9). Figure 2.2 shows how indicators reveal whether a company should be ranked with as Level 0, Level 1, Level 2, Level 3, or Level 4 in addressing climate change. This is followed by a discussion on how the tool has been applied and what information it has yielded.

Use of this Transition Pathway Initiative tool has grown significantly. It is now used in a major international investment effort called Climate Action 100+ to evaluate whether companies have established long-term quantitative emissions targets and are committed to using climate scenarios in their planning (Climate Action 100+ 2019, 21, 23), and investors can encourage companies to take such action if companies are not appearing to do so. Initial analysis using the tool has revealed that there is generally a largely even spread across Level 1 through Level 4 (Dietz, Byrne et al. 2019, 11). This indicates that the majority of companies are not currently integrating carbon risk considerations into management decisions, especially with respect

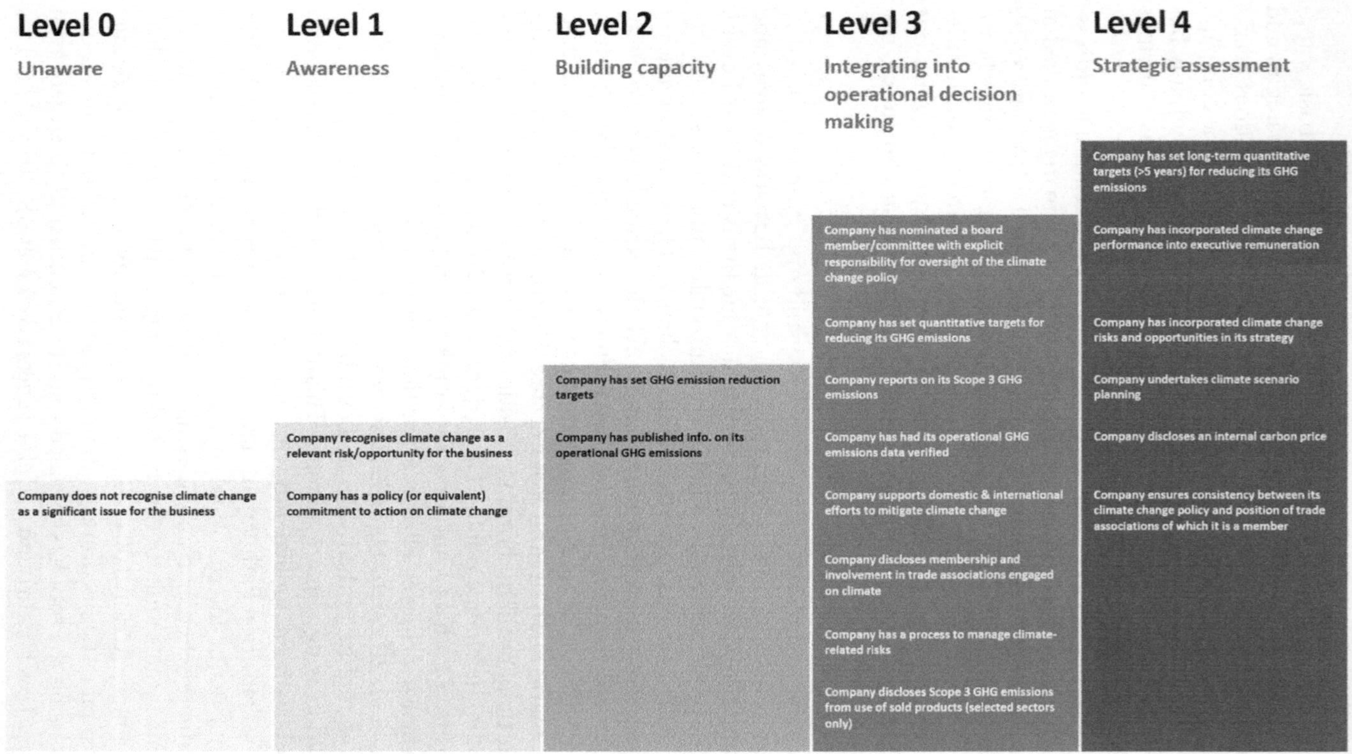

Figure 2.2 Transition Pathway Initiative Indicators and Rankings

Source: Transition Pathway Initiative

to recently developed Scope 3 emissions reporting and scenario analysis. However, more companies will likely receive higher level ratings as more companies integrate carbon risk considerations into management decisions, unless the Transition Pathway Initiative significantly modifies its indicators for each level.

2.3 Emerging efforts

At this time, reporting on carbon risk is largely voluntary. Even in jurisdictions where reporting is mandatory, regulators do not appear to be aggressively forcing companies to report or otherwise pursuing actions against companies for failing to appropriately report information pursuant to a reporting framework. This could change in the near future, however, as the largest financial markets, the United States, Europe, and China, are moving toward mandating reporting carbon risks in standard financial disclosures. With major financial markets having mandates for reporting carbon risks in standard financial disclosures, many international companies will be required to or will simply decide to report on carbon risk in their standard financial disclosures. It is also possible that a private cause of action could develop that would allow shareholders or other parties to pursue litigation against companies that altogether fail to report carbon risks or that do so inappropriately. This could also motivate widescale disclosures.

2.3.1 United States Securities and Exchange Commission requirements

The United States Securities and Exchange Commission currently requires companies to disclose material risks pursuant to the Securities Exchange Act of 1934, but there is not clear guidance as to what extent this entails disclosing environmental risks such as climate-related risks. This could change, though. The United States House Committee on Financial Services has been passing a series of bills that lay a framework for carbon risk disclosures, but counterpart bills are not being passed in the United States Senate.

One of the most notable bills is focused on Environmental, Social, and Corporate Governance (ESG) which is a set of factors that are viewed as central in addressing sustainable and ethical corporate practices. The ESG Disclosure Simplification Act of 2019 states that ESG metrics amount to de facto material risks for the purpose of disclosures under the Securities Exchange Act of 1934. Moreover, the Act requires companies to disclose in the notes section of financial filings how ESG metrics are being incorporated into business strategy (ESG Disclosure Simplification Act 2019, §2).

Another important bill is the Climate Risk Disclosure Act of 2019. The Climate Risk Disclosure Act of 2019 lays out specific climate change reporting requirements for a range of sectors, including the financial sector. The Act explains that companies have a duty to disclose climate-related financial risks and recommends doing so in annual reports to the Securities and Exchange Commission. According to the Act, it increases transparency to make standardized disclosures on material carbon risks that are useful to decision makers. Moreover, the Act explains that disclosures are critical in addressing climate change and help encourage a smooth transition to a low-carbon economy (Climate Risk Disclosure Act 2019, §3).

The Climate Risk Disclosure Act of 2019 then seeks to develop what would arguably be the most aggressive and advanced required reporting framework anywhere in the world. It does so by first adding provisions to Section 13 of the Securities and Exchange Act of 1934 as Section 13(s). These provisions require companies to report on how they identify, evaluate, and manage transition and physical risks (Climate Risk Disclosure Act 2019, §5). The Climate Risk Disclosure Act of 2019 then requires the Securities and Exchange Commission to issue rules identifying required disclosures under Section 13(s) of the Securities and Exchange Act of 1934. For example, these rules would involve reporting on Kyoto greenhouse gases and additional gases. In addition, these rules would require reporting on the specific fossil fuel securities being held. Further, these rules would involve reporting use of a 1.5 degrees Celsius scenario which reflects the large, immediate, and unprecedented global efforts necessary to mitigate greenhouse gas emissions to a level that is currently deemed desirable. Moreover, the rules would involve a discussion of short-, medium-, and long-term management strategies incorporating these metrics and other identified metrics (Climate Risk Disclosure Act 2019, §6). The Act also includes a backstop that becomes effective if the Securities and Exchange Commission does not issue rules identifying required disclosures. This backstop acknowledges that companies will be deemed in compliance with Section 13(s) of the Securities and Exchange Act of 1934 if their disclosures satisfy the TCFD recommendations (Climate Risk Disclosure Act 2019, §8).

2.3.2 Additional European disclosure requirements

Since 2018, the European Union has required listed companies, banks, and insurers with more than five hundred employees to make environmental disclosures in their annual reports. This can be done using international, national, or European reporting frameworks (EU 2014; EC n.d.a). Along these lines, the European Commission has been publishing voluntary guidance on environmental disclosures companies can choose to use in order to satisfy their disclosure requirements (EC 2017a, 1, 2017b). In 2019, the European Commission also published supplemental guidelines on reporting climate-related information that it states companies can use starting in 2020 (EC 2019a, 3, 2019b). In making environmental disclosures generally and climate-related disclosures specifically, the European Commission has emphasized that disclosures be material, fair, balanced, understandable, and comprehensive but concise (EC 2017a, 3, 2017b, 2019a, 1, 2019b).

A supporting document for the guidelines recognizes the TCFD recommendations as "authoritative guidance on the reporting of financially material climate-related information" and states that its new guidelines integrate the TCFD recommendations (EC 2019a, 2). As part of efforts connected with a European Green Deal, the European Commission has also launched a public consultation on disclosures (EC n.d.b).

Meanwhile, economies similar in size to France may start creating their own financial reporting frameworks, too. For instance, there have been multiple indications that mandatory disclosures are likely to be forthcoming in the United Kingdom. For instance, the Government of the United Kingdom recently published a green finance strategy, where it stated that "[t]he Government expects all listed companies and large asset owners to be disclosing in line with the TCFD recommendations by 2022" (HM Government 2019, 23). Then, only a few months later, the United Kingdom's Financial Conduct Authority proposed that listed United Kingdom companies will have to start disclosing their carbon risks next year. The framework for this reporting is not entirely clear, but the Financial Conduct Authority states that the expectation is for reporting to initially be done on a "comply or explain" basis (Binham and Hook 2019). This seems to be a similar strategy to that undertaken by France, allowing for the development of best practices while still pushing for disclosure.

2.3.3 Chinese disclosure requirements

For the past several years, the Chinese government has been exploring disclosure requirements. In 2017, the People's Bank of China started collaborating on a multiyear climate and environmental disclosure pilot with the Bank of England, the City of London, and the Principles for Responsible Investment (PRI), which oversees its own investor-focused PRI Reporting and Assessment Framework (PBOC 2019). The Hong Kong stock exchange is requiring that exchange-listed companies, with fiscal years beginning in July 2020, prepare a statement explaining how their boards evaluate ESG risks and materiality (HKEX 2020; Temple-West and Liu 2020). Other Chinese stock exchanges are expected to follow Hong Kong's lead, and the China Securities Regulatory Commission has announced its intention to implement broader action on this front (Temple-West and Liu 2020).

References

2 Degrees Investing Initiative. 2015. *Decree Implementing Article 173-VI of the French Law for the Energy Transition*. Paris: 2 Degrees Investing Initiative.

Binham, Caroline, and Leslie Hook. 2019. "Markets Watchdog Seeks to Force Companies to Disclose Climate Risk." *Financial Times*, October, 16. www.ft.com/content/837eea82-f02d-11e9-ad1e-4367d8281195.

California Department of Insurance. 2018. *Climate Risk Disclosure Survey Guidance Reporting Year 2018*. Sacramento: California Department of Insurance.

California Department of Insurance. n.d. "NAIC Climate Risk Disclosure Survey." Accessed July 25, 2020. www.insurance.ca.gov/0250-insurers/0300-insurers/0100-applications/ClimateSurvey/index.cfm.

CDP. n.d.a. "CDP." Accessed July 25, 2020. www.cdp.net/en.

CDP. n.d.b. "CDP Climate Change 2020 Questionnaire." Accessed July 25, 2020. https://guidance.cdp.net/en/tags?cid=13&ctype=theme&gettags=0&idtype=ThemeID&incchild=1&microsite=0&otype=Questionnaire&page=1&tgprompt=TG-124%2CTG-127%2CTG-125.

Climate Action 100+. 2019. "2019 Progress Report."

Climate Disclosure Standards Board and Sustainability Accounting Standards Board. 2019. *TCFD Implementation Guide: Using SASB Standards and the CDSB Framework to Enhance Climate-Related Financial Disclosures in Mainstream Reporting*. San Francisco: Climate Disclosure Standards Board and Sustainability Accounting Standards Board.

Climate Risk Disclosure Act of 2019, H.R.3623 (2019).

Dietz, Simon, Rhoda Byrne, Dan Gardiner, Valentin Jahn, Michal Nachmany, Jolien Noels, and Rory Sullivan. 2019. *TPI State of Transition Report 2019*. London: Transition Pathway Initiative.

Dietz, Simon, Valentin Jahn, Michal Nachmany, Jolien Noels, and Rory Sullivan. 2019. *Methodology and Indicators Report*. London: Transition Pathway Initiative.

ESG Disclosure Simplification Act of 2019, H.R.4329 (2019).

European Commission. 2017a. *Frequently Asked Questions: Guidelines on Disclosure of Non-Financial Information*. Brussels: European Commission.

European Commission. 2017b. *Guidelines on Non-Financial Reporting*. Brussels: European Commission.

European Commission. 2019a. *Frequently Asked Questions: Guidelines on Reporting Climate-Related Information*. Brussels: European Commission.

European Commission. 2019b. *Guidelines on Non-Financial Reporting: Supplement on Reporting Climate-Related Information*. Brussels: European Commission.

European Commission. n.d.a. "Non-Financial Reporting." Accessed July 25, 2020. https://ec.europa.eu/info/business-economy-euro/company-reporting-and-auditing/company-reporting/non-financial-reporting_en.

European Commission. n.d.b. "Non-Financial Reporting by Large Companies (Updated Rules)." Accessed July 25, 2020. https://ec.europa.eu/info/law/better-regulation/have-your-say/initiatives/12129-Revision-of-Non-Financial-Reporting-Directive/public-consultation.

European Union, Directive 2014/95/EU (2014).

French Energy Transition Law, 2015–992 (2015).

Global Reporting Initiative. n.d. "GRI Standards." Accessed July 25, 2020. www.globalreporting.org/standards.

Global Reporting Initiative and Global Sustainability Standards Board. 2016a. *GRI 102: General Disclosures*. Amsterdam: Global Reporting Initiative and Global Sustainability Standards Board.

Global Reporting Initiative and Global Sustainability Standards Board. 2016b. *GRI 103: Management Approach*. Amsterdam: Global Reporting Initiative and Global Sustainability Standards Board.

Global Reporting Initiative and Global Sustainability Standards Board. 2016c. *GRI 201: Economic Performance*. Amsterdam: Global Reporting Initiative and Global Sustainability Standards Board.

Global Reporting Initiative and Global Sustainability Standards Board. 2016d. *GRI 302: Energy*. Amsterdam: Global Reporting Initiative and Global Sustainability Standards Board.

Global Reporting Initiative and Global Sustainability Standards Board. 2016e. *GRI 305: Emissions*. Amsterdam: Global Reporting Initiative and Global Sustainability Standards Board.

Global Reporting Initiative and Global Sustainability Standards Board. 2016f. *GRI 415: Public Policy*. Amsterdam: Global Reporting Initiative and Global Sustainability Standards Board.

Global Reporting Initiative, Global Sustainability Standards Board, and CDP. 2017. *Linking GRI and CDP: How Are the GRI Sustainability Reporting Standards and CDP's 2017 Climate Change Questionnaire Aligned?* Amsterdam: Global Reporting Initiative, Global Sustainability Standards Board, and CDP.

Greenhouse Gas Protocol. n.d. "Greenhouse Gas Protocol." Accessed July 25, 2020. http://ghgprotocol.org/about-us.

Hong Kong Exchange. 2020. *How to Prepare an ESG Report*. Hong Kong: Hong Kong Exchange.

HM Government. 2019. *Green Finance Strategy*. London: HM Government.

International Association of Insurance Supervisors and Sustainable Insurance Forum. 2018. *Issues Paper on Climate Change Risks to the Insurance Sector*. Geneva: International Association of Insurance Supervisors and Sustainable Insurance Forum.

International Association of Insurance Supervisors and Sustainable Insurance Forum. 2020. *Issues Paper on the Implementation of the Recommendations of the Task Force on Climate-related Financial Disclosures*. Geneva: International Association of Insurance Supervisors and Sustainable Insurance Forum.

Mason, Amy, Will Martindale, Alyssa Heath, and Sagarika Chatterjee. 2016. *French Energy Transition Law: Global Investor Briefing*. London: Principles for Responsible Investment.

Messervy, Max, and Cynthia McHale. 2016. *Insurer Climate Risk Disclosure Survey Report & Scorecard: 2016 Findings & Recommendations*. Boston: Ceres.

Mills, Evan, Ted Lamm, Sadaf Sukhia, Ethan Elkind, and Aaron Ezroj. 2018. *Trial by Fire: Managing Climate Risks Facing Insurance in the Golden State*. Sacramento: California Department of Insurance.

National Association of Insurance Commissioners. n.d. "Climate Risk Disclosure." Accessed January 10, 2019. https://content.naic.org/cipr_topics/topic_climate_risk_disclosure.htm.

People's Bank of China, Bank of England, City of London, and Principles for Responsible Investment. 2019. *UK-China Climate and Environmental Information Disclosure Pilot*. Beijing: People's Bank of China, Bank of England, City of London, and Principles for Responsible Investment.

Securities Exchange Act of 1934, 15 U.S.C. § 78a et seq (1934).

Task Force on Climate-related Financial Disclosures. 2017. *Recommendations of the Task Force on Climate-related Financial Disclosures*. Basel: Task Force on Climate-related Financial Disclosures.

Temple-West, Patrick, and Nian Liu. 2020. "Chinese Companies Get to Grips with Tougher ESG Disclosures." *Financial Times*, January 13. www.ft.com/content/b06291aa-3251-11ea-9703-eea0cae3f0de.

3 Carbon footprint calculations

Carbon footprint calculations can identify carbon risks with particular companies, especially within supply chains.

With an increasing scientific consensus that rising concentrations of greenhouse gases were likely attributable to human activity and that the gases could potentially destabilize the world's climate system, delegates from nearly two hundred countries gathered in Kyoto for the 1997 United Nations Climate Change Conference. To address these concerns, delegates drafted an international climate change treaty known as the Kyoto Protocol. This treaty extended United Nations Framework Convention on Climate Change commitments to reduce atmospheric greenhouse gas concentrations to levels that would prevent dangerous interference with the word's climate system (Kyoto Protocol 1997, Art. 3). The treaty applies to six greenhouse gases that vary in their global warming capacity and prevalence: carbon dioxide, methane, nitrous oxide, hydrofluorocarbons, perfluorocarbons, and sulfur hexafluoride (Kyoto Protocol 1997, Annex A). Since this time, nearly two hundred countries have adopted the Kyoto Protocol (UN 2020).

The Kyoto Protocol highlighted the need for reducing greenhouse gas emissions across a broad range of companies. Therefore, following the Kyoto Protocol, companies started calculating their carbon footprints, which is how much greenhouse gas they emit into the atmosphere as a direct or indirect result of their operations. Through such an analysis, companies are able to compare the extent of their emissions to peers within their sector of the economy and identify potential opportunities for reducing emissions.

3.1 Quantifying emissions

The Greenhouse Gas Protocol developed a reporting methodology to quantify emissions: the GHG Protocol Corporate Standard. Most companies currently use this methodology to record and report their carbon footprints. The GHG Protocol Corporate Standard explains accounting and reporting procedures for the six Kyoto Protocol greenhouse gases, dividing emissions into three scopes, Scope 1, Scope 2, and Scope 3. GHG Protocol Corporate Standard requires reporting Scope 1 and Scope 2 emissions at a minimum, with reporting Scope 3 emissions being viewed as optional (GHG Protocol 2015, 3, 25). Table 3.1 provides a brief overview of the different emissions scopes under the GHG Protocol Corporate Standard.

3.1.1 Emissions scopes

Scope 1 emissions are companies' direct greenhouse gas emissions from owned or controlled sources (GHG Protocol 2015, 25). Some of the larger sources of such emissions include, for instance, emissions are from companies' power facilities or vehicles. Therefore, utilities that generate electricity and transportation companies such as airlines are likely to have significant Scope 1 emissions.

Scope 2 emissions are the indirect greenhouse gas emissions from the electricity that companies purchase. Scope 2 emissions physically occur at the facility where electricity is generated, such as a local power plant (GHG Protocol 2015, 25). Commercial buildings and manufacturing companies that consume significant electricity, such as concrete producers, are likely to have significant Scope 2 emissions.

Scope 3 emissions are the indirect emissions that occur in the value chain, including both upstream and downstream emissions. Upstream emissions include the extraction, production, and transportation of purchased materials. Downstream emissions include the use of sold products and services. Scope 3 emissions occur as the result of companies' activities but are from sources that are not owned or controlled by companies (GHG Protocol 2015, 25). For certain types of companies, Scope 3 emissions can be responsible for significant greenhouse gas emissions, significantly exceeding direct emissions (Trucost and S&P Global 2019, 1).

Tracking Scope 3 emissions is extremely difficult, because doing so requires making assumptions about the actions of sources that are not owned or controlled by companies, and there is likely to be significant variation based on such assumptions. Therefore, the GHG Protocol has published a separate GHG Protocol Scope 3 Standard to supplement the GHG Protocol Corporate Standard. The GHG Protocol Scope 3 Standard provides details on

Table 3.1 Emissions Scopes

Scope	*Description*
Scope 1 Emissions	Direct emissions from owned or controlled sources
Scope 2 Emissions	Indirect emissions from the purchase of electricity
Scope 3 Emissions	Indirect emissions that occur in the value chain, including both upstream and downstream emissions

where to obtain emissions data for specific types of activities and how to assess reliability for the data (GHG Protocol 2011).

The GHG Protocol Corporate Standard provides guidance on recording greenhouse gas emissions. Guidance on recording greenhouse gas emissions details the need for assessments made at the facility level and a discussion of any observed discrepancies. According to GHG Protocol Corporate Standard, companies must obtain and summarize data from facilities that could be located in different countries or be in different business divisions. Facility reports should include a brief description of the emission sources with a discussion on why specific sources were either excluded or included. The reports should further draw comparisons to prior years, identifying any trends or changes that could have impacted results (GHG Protocol 2015, 40–47).

Guidance on reporting greenhouse gas emission includes metrics that it explains should be relevant, complete, consistent, transparent, and accurate. Reports should include emissions data for all six covered greenhouse gases in metric tons and in tons of carbon dioxide equivalent. Emissions data should be provided separately for each scope. There should be a discussion on the scope of the covered sources and exclusions. Reports should identify significant changes. There should also be a discussion of the methodologies used to measure emissions, especially in relation to the base year. Optional items include greenhouse gas verification statements and a description of greenhouse gas reduction efforts (GHG Protocol 2015, 62–64).

3.1.2 Global warming potential

Table 3.2 shows how the six Kyoto Protocol greenhouse gases compare to one another with respect to global warming capacity. Global warming capacity is typically measured in tons of carbon dioxide. This allows for aggregation of the greenhouse gas emissions associated with different gases. However, there is not scientific consensus over precise numbers, given that chemical compounds perform differently under different atmospheric conditions and timeframes (GHG Protocol 2019). Moreover, hydrofluorocarbons and perfluorocarbons are a set of gases with varying chemical compositions

Table 3.2 Global Warming Potential

Greenhouse Gas	*Global Warming Potential*
Carbon Dioxide	1
Methane	21–28
Nitrous Oxide	265–310
Hydrofluorocarbons	Up to 14,800
Perfluorocarbons	Up to 12,200
Sulfur Hexafluoride	22,800 to 23,900

rather than a single distinct chemical composition (GHG Protocol 2019, 2–3). Therefore, it is important for any reporting party to explain their methodology for making calculations and, in some cases, a separate report for each type of gas is preferable.

3.2 Aggregating emissions

There are also efforts to encourage large investors to aggregate the carbon footprints of companies that they are invested in or otherwise quantify the carbon intensity of their investments portfolios as a whole. The most established of these efforts is the Montreal Carbon Pledge. In 2014, the Pledge was launched in Montreal with the support of PRI and the United Nations. PRI currently oversees implementation of the Pledge. More than one hundred investors, managing more than ten trillion dollars in assets, have already voluntarily signed the Pledge (Montreal Carbon Pledge n.d.).

When signing the Montreal Carbon Pledge, an investor agrees to measure and publicly disclose their investment portfolio's carbon footprint. A portfolio's carbon footprint is the sum of each portfolio company's emissions proportional to the amount of stock that the company makes up in the portfolio. The Pledge does not provide a specific methodology for aggregating the carbon footprints of portfolio companies, nor does it specify whether calculations should include both direct and indirect emissions. However, the United Nations and the Greenhouse Gas Protocol have released guidelines on good practice that are being used. The Pledge website further explains that PRI encourages all investors to implement the TCFD's disclosure guidance, which provides additional direction (Montreal Carbon Pledge n.d.).

While signing the Montreal Carbon Pledge does not obligate an investor to take any additional action, the PRI further encourages Pledge signatories to further address their portfolios' carbon footprints. Such means for addressing a carbon footprint could include, for instance, regularly tracking emissions to ensure that they are understood, engaging with investee companies on their emissions disclosure and management, incorporating environmental considerations into investment decisions, decreasing exposure to carbon-intense investments, and investing in low-carbon solutions such as improved energy efficiency or substitution of fossil-fuel–based electricity with renewably generated electricity. Pledge signatories can also join further collaborations where they can collectively implement these means for addressing the carbon footprints of portfolios, such as the Portfolio Decarbonization Coalition. This is a multistakeholder effort seeking to reduce greenhouse gas emissions by mobilizing institutional investors to gradually decarbonize the investment portfolios. According to the Pledge website, a portfolio's carbon footprint is a useful quantitative tool that can help an investor develop a broad climate change strategy (Montreal Carbon Pledge n.d.).

3.3 Evaluating investments

The TCFD also advised on advanced exposure metrics for aggregating carbon footprints within an investment portfolio. This included a discussion of several metrics where Scope 1 and Scope 2 emissions were allocated to investors based on an equity ownership approach, and another metric focused on examining a portfolio's exposure to sectors and industries considered the most greenhouse gas emission intensive. The TCFD highlighted the weighted average carbon intensity metric shown in Table 3.3 as the "[m] etric recommended by the Task Force" (TCFD 2017, 43–44).

Consultants are currently marketing services for applying advanced exposure metrics. The consulting firms can obtain company carbon footprint information from an inventory, such as that maintained by CDP, or they can generate their own carbon footprint information for companies. When using sector greenhouse gas data to derive exposure metrics, doing so is referred to as "top-down" greenhouse gas analysis. When instead aggregating individual company greenhouse data to derive exposure metrics, doing so is referred to as "bottom-up" analysis (Trucost and S&P Global 2019, 3).

Nevertheless, attempts to calculate the carbon footprints of investment portfolios have encountered some complications, such as the following. First, there has been a lot of variation observed in calculating carbon footprints, especially in calculating Scope 3 emissions. If multiple consultants calculated the carbon footprint for a particular company, it would not be uncommon for them reach very different results. Aggregating carbon footprints of companies across an investment portfolio involves numerous carbon footprint calculations for a number of companies. Therefore, it is presumably even more likely to see variation among aggregate calculations for entire portfolios. Second, companies are generally only compared to other companies in the same sector because there is significant variation in carbon footprints between sectors. For instance, carbon footprint metrics tend to calculate a low carbon footprint for companies in the service sectors and a high carbon footprint for companies in the energy, manufacturing, and transportation sectors. It could be a risky financial decision for companies to try to decrease the carbon footprints of their investment portfolios by trying to divest from entire sectors or heavily favor certain sectors.

Table 3.3 Weighted Average Carbon Intensity

Description	Portfolio's exposure to carbon-intensive companies, expressed in tons CO_2e/\$M revenue.
Methodology	Scope 1 and Scope 2 GHG emissions are allocated based on portfolio weights.
Formula	(Current Value of Investment/Current Portfolio Value) × (Issuer's Scope 1 and Scope 2 GHG Emissions/Issuer's \$M Revenue) (TCFD 2017, 43–44)

Because of these complications, calculating the carbon footprints of investment portfolios may not yield the most valuable or actionable information. At a minimum, it would be beneficial to have more available, reliable, and comparable granular greenhouse gas data.

References

Greenhouse Gas Protocol. 2011. *Greenhouse Gas Protocol, Corporate Value Chain (Scope 3) Accounting and Reporting Standard*. Washington, DC: World Resources Institute and World Business Council for Sustainable Development.

Greenhouse Gas Protocol. 2015. *The Greenhouse Gas Protocol, A Corporate Accounting and Reporting Standard* (Revised Edition). Washington, DC: World Business Council for Sustainable Development and World Resources Institute.

Greenhouse Gas Protocol. 2019. *Global Warming Potential Values*. Washington, DC: World Resources Institute.

Kyoto Protocol to the United Nations Framework Convention on Climate Change, December 10, 1997, 2303 U.N.T.S. 162.

The Montreal Carbon Pledge. n.d. "The Montreal Carbon Pledge." Accessed July 25, 2020. https://montrealpledge.org.

Task Force on Climate-related Financial Disclosures. 2017. *Implementing the Recommendations of the Task Force on Climate-related Financial Disclosures*. Basel: Task Force on Climate-related Financial Disclosures.

Trucost ESG Analysis and S&P Global. 2019. *Trucost Scope 3 Data: Analysis Across the Value Chain*. London: Trucost ESG Analysis and S&P Global.

United Nations Framework Convention on Climate Change. n.d. "The Kyoto Protocol – Status of Ratification." Accessed July 25, 2020. https://unfccc.int/process/the-kyoto-protocol/status-of-ratification.

4 Brown taxonomies

Brown taxonomies can help investors identify current carbon risks by classifying fossil fuel assets in a systematic manner.

Because of advancements with monitoring greenhouse gases and determining their global warming potential, it has become possible to estimate how much greenhouse gas could be emitted in the atmosphere before exceeding the Paris Agreement temperature objectives, although admittedly there is significant room for disagreement on the precision of the estimates, as there are numerous factors that need to be taken into account to make such estimates. Moreover, because of these same advancements, it is possible to project how different sectors of the economy and their associated greenhouse gas emissions could achieve or fail to achieve the greenhouse gas budgets needed to meet the Paris Agreement temperature targets. Again, there could be disagreement on the precision of estimates with different sectors of the economy, too. There are also various ways that energy can be deployed to achieve the same objectives. Nevertheless, despite the risk-management challenge, this kind of analysis has provided a framework for tracking investment activities for alignment with the Paris Agreement.

4.1 Carbon bubble hypothesis

In 2011, Carbon Tracker published its report entitled *Unburnable Carbon* that introduced the "carbon bubble" hypothesis. This hypothesis warned about an inflating carbon bubble if more capital is invested in carbon emission–producing assets, which could crash in value. According to the carbon bubble hypothesis, the desire to restrict climate change creates a carbon budget of the amount of emissions that the world can afford. And, according to estimates, such a budget only allows for the use of a fraction of the world's fossil fuel reserves, making the remaining vast majority of reserves unusable or stranded (Carbon Tracker 2011, 6).

Carbon Tracker explained that the Potsdam Climate Institute calculated that in order to appropriately reduce the chance of exceeding 2 degrees

Celsius, there was a limit on the amount of emissions that could be released into the atmosphere between 2000 and 2050. Even less could be emitted if it was desired to limit temperature increases to 1.5 degrees Celsius. At the same time, using corresponding emissions factors for coal, oil, and gas, the Potsdam Climate Institute calculated the total foreseeable emissions from burning the earth's identified coal, oil, and gas reserves. This revealed that burning all of these reserves would result in carbon being released into the atmosphere far in excess of the amount of emissions that could be released into the atmosphere while still likely achieving 2 degrees Celsius or lower targets by 2050. According to Carbon Tracker, at most, one-fifth of the Earth's reserves could be used by 2050 if the world wanted to reduce the chance of exceeding 2 degrees Celsius to twenty percent (Carbon Tracker 2011, 6). Figure 4.1 details the carbon bubble hypothesis.

Both international environmental and financial regulators have acknowledged the credibility of the carbon bubble hypothesis. Similar to Carbon Tracker, the IPCC concluded that in order to limit global temperature rise to 2 degrees Celsius above preindustrial levels, only between one-fifth and one-third of the world's proven oil, gas, and coal reserves could be burned (Carney 2015; IPCC 2014; Hill and Martinez-Diaz 2019, 56–58). Governor Carney referenced the carbon bubble hypothesis in his *Breaking the Tragedy of the Horizon* speech. He said that financial markets were not appropriately valuing fossil fuel assets and other assets tied to them. As Governor Carney explained, the desire to restrict climate change creates a carbon "budget" of the amount of emissions that the world can "afford." And, according to estimates, such a budget only allows for the use of a fraction of the world's fossil fuel reserves, making the remaining vast majority

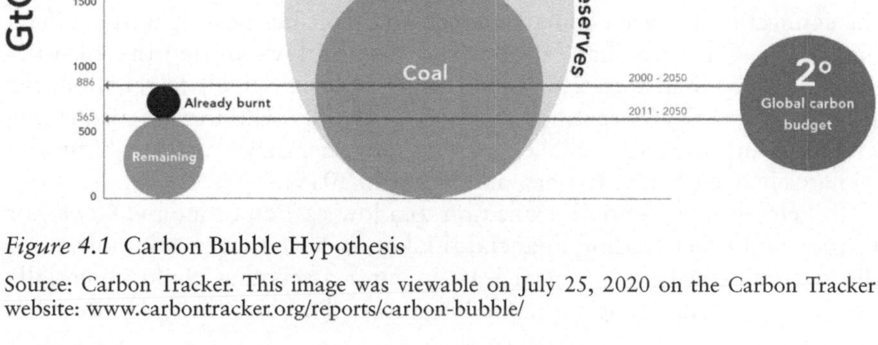

Figure 4.1 Carbon Bubble Hypothesis

Source: Carbon Tracker. This image was viewable on July 25, 2020 on the Carbon Tracker website: www.carbontracker.org/reports/carbon-bubble/

of reserves unusable or "stranded" (Carney 2015). However, assets tied to these reserves are being valued as if all the reserves can be used. A quick realization that these reserves are unburnable and therefore in fact stranded could lead to a market shock, that is, what some have equated to the bursting of a "carbon bubble" (Carney 2015; Carney, Villeroy de Galhau, and Elderson 2019). The Dutch Bank also discussed the carbon bubble hypothesis in a comprehensive climate report (Regelink et al. 2017, 12). While there is debate over how significant of an impact a reevaluation of carbon assets could have and what assets in particular are most at risk of becoming stranded in a transition to a low-carbon economy (Carbon Tracker 2019), it is becoming generally accepted within the global financial community that certain assets may become stranded without significant advancements in carbon storage or other technologies limiting emissions from the burning of fossil fuels.

In fact, shifts in the global economy are already observable. In the last several years, coal has experienced major losses. The Dow Jones United States Coal Index decreased over ninety percent in value from April 2011 to June 2017 (S&P Global n.d.). Dozens of coal companies filed for bankruptcy between 2012 and 2015 (Kuykendall 2015). In the United States, the percentage of electricity generated from coal has dropped significantly (Levisohn 2017), and coal production has plummeted (Krauss 2016). In 2020, for the first time, coal will likely have produced less electricity in the United States than renewable energy (Plumer 2020). Moreover, major banks are refusing to invest in new coal infrastructure (Buckley 2019), and major insurance companies are restricting underwriting for coal infrastructure (AXA 2020, 30). Financial support in Asia may even be stemmed, with Japan's environment minister vowing to slash financing that has supported construction of coal infrastructure in India, Vietnam, and Indonesia (Harding 2020).

Other fossil fuel assets also appear likely to face challenges in the near future. Major economies, such as the United Kingdom, France, the Netherlands, California, China, and Singapore, are planning to ban the sale of gasoline and diesel-powered vehicles in the coming decades (Petroff 2017; Newsom 2020; BBC 2017; Aravindan and Geddie 2020). Shareholders are suing oil companies over their failure to disclose their knowledge on the connection between climate change and their business operations (Savage 2019). Oil prices have dropped to historic lows during the COVID-19 pandemic (Gladstone 2020), and while prices are likely to rebound, the price drop has clearly revealed the asset's volatility. Utilities are also facing mounting infrastructure challenges that they themselves are attributing to climate change (Brody, Rogers, and Siccardo 2019).

To help ensure a smooth transition to a low-carbon economy, Governor Carney and other leading financial thinkers believe that it is important to divest from, engage with, or at least monitor assets that could potentially become stranded. Along the same lines, to replace lost energy production, there needs to be more investment in sustainable or green energy. Indeed, there have been a number of efforts to track specific equity and fixed income investments to determine whether they are connecting with sectors of the economy in a manner that would ensure achievement of sustainability ambitions, such as the Paris Agreement objectives.

4.2 Classifications

As fossil fuel assets could potentially become stranded, the financial community is increasingly monitoring these assets and in some cases is encouraging either divestment or engagement as a means of addressing carbon risk. To do so the government and nonprofit organizations have developed classification systems for identifying fossil fuel investments which can be viewed as brown taxonomies or precursors to such if the government or nonprofit organizations managing them do not formally or informally view them as brown taxonomies. The development and application of these classification systems has been made possible through already established financial practices for labeling securities with unique numerical identifiers (CUSIP Global Services n.d.; ISIN Organization n.d.). These unique numerical identifiers reflect the issuer of securities and whether they are debt or equity. The numerical identifiers have also allowed third-party data product providers to sell data products that can provide additional information for specific numerical identifies, including information on the sectors, industry groups, and subindustry groups that specific securities fit into (MSCI n.d.). There are now also third-party data providers, such as Trucost, that provide information on the percentage of revenue that companies are generating from fossil fuel enterprises and the percentage of electricity that utilities are generating from fossil fuel sources. Together, the numerical identifiers and third-party data products can be used to run an automated analysis across an investment portfolio to quickly identify the prevalence of fossil fuel investments.

Monitoring greenhouse gas emissions is helpful in informing a company how to act in reducing its own emissions. However, there are challenges when aggregating this information across an investment portfolio or communicating specific actions that can be taken to decision makers based solely on this information. Tracking investments using security information, on the other hand, either in combination with or separately from greenhouse gas emissions, allows for easier aggregation of data and more specific guidance that can more easily be communicated for investment decisions.

4.2.1 California pension funds

California has been a leader in advocating for divestment from what is believed to be one of the assets most as risk of becoming stranded: thermal coal. This is the type of coal that is burned to generate electricity rather than metallurgic coal that is used to make steel. As it is generally accepted that coal power plants will have to be retired in order to meet global emissions targets (Yanguas et al. 2019), and coal plants are now closing as countries such as the United Kingdom, France, the Netherlands, and Germany move forward with phasing out coal power (Johnson 2020; Reuters 2019; Netherlands 2019; BBC 2020), there is a decreasing need for this fuel source under most future energy projections.

California's initial push to limit investment in thermal coal was through state legislation signed into law by Governor Jerry Brown in 2015. This legislation directed the state's two largest pension funds, the California Public Employees' Retirement System (CalPERS), managing over three hundred billion dollars in assets (CalPERS n.d.), and the California State Teachers' Retirement System (CalSTRS), managing over two hundred billion dollars in assets (CalSTRS 2020), to completely divest from thermal coal companies and not make future investments in those companies unless the companies showed that they were transitioning to a clean-energy future (Public Divestiture of Thermal Coal Companies 2015). The legislation, however, did not require the CalPERS and CalSTRS boards to take any action unless the boards determined in good faith that such action was consistent with the board's fiduciary responsibilities (Public Divestiture of Thermal Coal Companies 2015, §1).

The state legislation added a section to California's Government Code, reading in part:

> The combustion of coal resources is the single largest contributor to global climate change in the United States.
>
> . . .
>
> Climate change affects all parts of the California economy and environment, and the Legislature has adopted numerous laws to mitigate greenhouse gas emissions and to adapt to a changing climate.
>
> . . .
>
> The purpose of this section is to require the Public Employees' Retirement System and the State Teacher's Retirement System, consistent with, and not in violation of, their fiduciary responsibilities, to divest their holding of thermal coal power as one part of the state's broader efforts to decarbonize the California economy and to transition to clean, pollution free energy sources.
>
> (Public Divestiture of Thermal Coal Companies 2015, §1)

Thermal coal companies were identified as those companies that derive fifty percent or more of their revenue from the mining of thermal coal (Public Divestiture of Thermal Coal Companies 2015, §1). This pension fund classification system, which has a fairly straightforward coal classification, is detailed Table 4.1. This is followed by a discussion on how both the CalPERS and CalSTRS boards acted with respect to this classification.

Table 4.1 California Pension Funds – Brown Classification

Classification	*Description*
Thermal Coal Enterprises – Percentage	Companies that derive fifty percent or more of their revenue from the mining thermal coal

Both the CalPERS and CalSTRS boards subsequently found that divestment was consistent with their fiduciary duties and voted to take such action. The CalPERS investment committee made its decision to divest in closed session. Although minutes are not available for that session, agenda materials are available and reveal actions taken by staff. These materials show that staff engaged seventeen coal companies. Fourteen offered no transition plan away from coal, and three offered a transition plan to a cleaner energy model. The CalPERS investment committee decided to remain invested with the three companies that offered a transition plan and divested from those that did not (CalPERS 2017, 3).

The CalSTRS board publicly voted to divest from thermal coal companies (CalSTRS 2017). Board members made the following statements when doing so. California State Controller Betty Yee stated:

> As fiduciaries working to maximize the retirement benefits that educators have earned during their careers, we know it is clear that the liability of any investment in thermal coal is too great[.] . . . The regulatory risk and environment impacts climate change places on the fund far outweighs the ability of thermal coal companies to continue to create long-term value.
>
> (CalSTRS 2017)

California State Treasurer John Chiang further stated:

> Coal is the fuel of America's past. At a time when clean-energy jobs far outpace those of the dying coal industry, why should we continue to invest in a product that pollutes the air, sickens our children and contributes to global warming[.]
>
> (CalSTRS 2017)

CalPERS ultimately divested approximately fourteen million dollars from thermal coal assets (CalPERS 2017, 3), and CalSTRS ultimately divested approximately ten million dollars from thermal coal assets (CalSTRS 2017). Both amounts were relatively small, with a total of over five hundred billion dollars that they collectively manage, most of that being publicly traded stock (CalPERS 2020; CalSTRS 2020). The California pension fund effort mirrors similar efforts taken by smaller government pension funds elsewhere in the United States and Europe. Nordic countries such as the Norwegian Government Pension Fund were the first to divest, with guidance from a European-based NGO with divestment expertise, Urgewald (n.d.b).

4.2.2 Climate Risk Carbon Initiative

A second push for thermal coal divestment came from California's insurance regulator, the California Department of Insurance. Unlike elsewhere in the world, insurance is regulated at the subnational level in the United States (McCarran-Ferguson Act 1945). This provides larger states with unique regulatory authority akin to countries in other parts of the world. The California Department of Insurance, as the largest insurance regulator in the United States and one of the largest in the world, asked all insurance companies operating in California to voluntarily divest from thermal coal. In doing so, then California Insurance Commissioner Dave Jones wrote to insurance companies stating the following in an effort he described as the California Department of Insurance's Climate Risk Carbon Initiative:

> I am writing to request that your company voluntarily divest from any investments it may have in thermal coal.
>
> . . .
>
> At some point nations and states may dramatically restrict the burning of carbon. At that point, investments in coal mines, in oil and gas wells, in companies that extract coal, oil or natural gas, in companies that transport coal, oil and gas, in utilities that rely on coal, oil or gas, among others, could drop dramatically in value. Before that happens it is important for insurance companies and insurance regulators to understand the scope of these investments by insurance companies and to take steps to mitigate potential financial risks.
>
> Divestment from thermal coal in particular will help protect insurance companies from holding an investment currently dropping in value, and which is likely to suffer substantial additional decline in value during a transition to a reduced carbon economy, and run the risk of becoming a "stranded asset."
>
> (Jones 2016)

Moreover, then-Commissioner Jones asked all insurance companies to answer questions on any plans to divest from thermal coal and asked larger insurance companies to disclose a list of their fossil fuel investments (Jones 2016). Guidance for disclosing fossil fuel investments explained that the fossil fuel investments that should be disclosed included any investments meeting the following criteria: investments in companies generating over thirty percent of their revenue from thermal coal, investments in utilities generating over thirty percent of their electricity from the burning of thermal coal, investments in companies generating over fifty percent of their revenue from oil and gas combined, and investments in utilities generating over fifty percent of their electricity from the burning of coal, oil, and gas combined (CDI 2016, 4, 7–8). Table 4.2 details the classification system.

Using responses from over six hundred of the largest insurance companies operating in California and supplemental data provided by a consultant, the California Department of Insurance built a searchable database of fossil fuel investments being held by all insurance companies operating in California. The database can be searched for a specific company, the total assets that the company has under management, the amount of those assets that are thermal coal or other fossil fuel investments, and whether the company has any plans to divest from thermal coal. Clicking on certain categories also provides more detailed information, such as information on the types of investments being held and the names of those investments. Queries can also be run that reveal outliers, such as those companies holding the greatest amount of categorized thermal coal investments. Certain companies have much larger concentrations of such assets than other companies (CDI n.d.).

The California Department of Insurance has continued to supplement its database. However, in subsequent years, the Department worked with a consultant to obtain all financial data from annual filings rather than seeking responses from the largest insurance companies operating in California and supplementing that data. Information for subsequent years can be accessed through the same searchable database by selecting the year for which data is desired (CDI n.d.).

The California Department of Insurance also provides aggregate fossil fuel numbers and charts comparing annual data. This data reveals that insurers doing business in California have over five hundred billion dollars, or approximately eleven percent, of their assets under management invested in fossil fuels. Of this over five hundred billion dollars, over a hundred billion is invested in thermal coal. The majority is in utility investments. The minority, approximately twenty-one billion dollars, is invested in thermal coal enterprises (CDI n.d.).

Table 4.2 Climate Risk Carbon Initiative – Brown Classification

Classification	*Description*
Thermal Coal Enterprises – Percentage	Companies that derive over thirty percent of their revenue from thermal coal
Utility Coal Power Generation	Utilities that generate over thirty percent of their electricity from burning thermal coal
Fossil Fuel Enterprises Excluding Coal – Percentage	Companies that derive over fifty percent of their revenue from oil and gas combined
Utility Fossil Fuel Power Generation	Utilities that generate over fifty percent of their electricity from burning thermal coal, oil, and gas combined

The amount of total fossil fuel investments and thermal coal investments has stayed relatively flat across the years analyzed by the Department (CDI n.d.). Nevertheless, there is reason to believe that divestment is taking place. To begin with, this is because the price of these assets has increased slightly since former Commissioner Jones first called for divestment. For several years, the Dow Jones United States Coal Index decreased dramatically, but then increased slightly (S&P Global n.d.). Following a dramatic price decrease, the slight price increase could make the aggregate coal numbers look similar across years even if less of that asset is being held in subsequent years. Moreover, another reason to believe that divestment is taking place is because a supplemental survey conducted in 2018 showed that the number of insurers divesting from thermal coal has increased since the initial call for divestment and the survey conducted in connection with that call (CDI 2018).

4.2.3 Global Coal Exit List

On their own, companies are also moving forward with divesting from thermal coal, asserting both financial and environmental reasoning. While some of these companies originally classified thermal coal the same as or similar to the classifications used by the California Department of Insurance, some of these companies have since expanded that classification to also include companies that generate significant revenue from coal in the aggregate and utilities that are expanding their ability to generate electricity from the burning of thermal coal. Major international companies are reporting thermal coal divestment in their TCFD and French Energy Transition Law reports. Divestment classifications are becoming more aggressive, too. For instance, in a TCFD and French Energy Transition Law–aligned report, one of the world's largest insurers, AXA, explained its thermal coal divestment strategy as follows.

> Carbon emissions will require significant curbing in order to reduce the risk of climate change, which may place business constraints on carbon-intensive industries, leaving some assets "stranded", which in turn may lead to reduced valuations. Current valuation models may not account for such risks adequately. This is why AXA had decided in 2015 to divest [more than 500 million dollars] from the coal industry, by targeting coal mining and coal-based electric utilities which derive over 50% of their revenues from coal.
>
> (AXA 2018, 15)

AXA then explained that in 2017, it decided to increase its thermal coal divestment to reach billions of dollars by expanding its coal exclusion criteria by several means. To begin with, the criteria were expanded to include enterprises deriving over thirty percent of their revenues from coal and utilities generating over thirty percent of their electricity from the burning of coal, which is more inclusive than its original fifty percent thresholds. According to AXA, this captures long-term financial risks related to "stranded assets." Further, AXA expanded its coal exclusion criteria to include companies with significant coal-based power "expansion plans." According to AXA, such companies are constructing new coal power plants that are locking economies into coal power for decades, which clearly contradicts the Paris Agreement. Moreover, AXA expanded its coal exclusion criteria to include mining companies that annually produce large quantities of coal (AXA 2018, 15–16). AXA has continued to expand its coal criteria (AXA 2020, 26, 30).

Some companies that are divesting from coal may hire their own consultants to identify coal securities in their investment portfolios. At least some companies that are divesting from thermal coal use the Global Coal Exit List, though, or related resources to identify thermal coal investments, including AXA and other major reinsurers such as Swiss Re and Zurich (Urgewald n.d.b). The Global Coal Exit List, constructed by Urgewald, contains hundreds of companies identified as being connected with the coal industry. Along with information identifying each company and some background information on that company's activities if they are in the power sector, there is data corresponding to categories of activities that asset owners can use to identify a company as a coal investment. These categories include the revenue and power thresholds used by the California Department of Insurance to identify thermal coal, companies with over thirty percent of their revenue derived from coal, and utilities generating over thirty percent of their electricity from coal. The categories also include aggregate numbers, such as coal companies producing large quantities of coal and utilities with the capacity to generate large amounts of electricity from burning coal. There is also information on companies expanding their thermal coal activities. This includes companies significantly expanding their thermal coal mining or infrastructure development and utilities that are planning substantial increases in their capacity to generate electricity from coal, for example, by over three hundred megawatts (Urgewald n.d.a). Table 4.3 details the classification system.

Urgewald hosts a public version of its Global Coal Exit List on its website. Asset owners can also obtain a list with security identification numbers from Urgewald. This will further facilitate matching securities within asset owners' portfolios with thermal coal designations in case the asset owner has difficult identifying securities from the companies identified in the public version of the Global Coal Exit List (Urgewald n.d.a).

Table 4.3 Global Coal Exit List – Brown Classification

Classification	*Description*
Thermal Coal Enterprises – Percentage	Companies that derive a certain percentage of their revenue from thermal coal
Utility Coal Power Generation – Percentage	Utilities that generate a certain percentage of their electricity from burning thermal coal
Thermal Coal Enterprise – Aggregate	Companies that derive a certain amount of revenue from thermal coal
Utility Coal Power Generation – Aggregate	Utilities that generate a certain amount of electricity from burning thermal coal
Thermal Coal Expansion	Companies significantly expanding their thermal coal mining or infrastructure development and utilities that are planning substantial increases in their capacity to generate electricity from coal

4.2.4 Additional taxonomies

In addition to divesting from thermal coal assets, there are also efforts to divest from or otherwise restrict investments in specific subset of oil assets, tar, or oil sands. Oil sands potentially pose unique regulatory and environmental risks that could impact their price in the long run. Depending on reserves and processing capabilities, they may also be more expensive to extract and refine than other oil sources, making them more exposed to price decreases. For example, the San Francisco State University Foundation, which manages the university's endowment, voted to make no new investments in "direct ownership of companies with significant exposure to production or use of coal and tar sands" (Examiner Staff 2014). A number of companies have also stated that they will divest from and no longer invest in oil sands. In its TCFD and French Energy Transition Law–aligned report, in addition to explaining its coal divestment strategy, AXA explained its oils sands divestment strategy. This included divesting from the main oil sands producers, which it defined as producers with at least thirty percent of their reserves relying on oil sands, and from the primary companies involved with the development of oil sands pipelines. AXA explained that it was taking these steps because oil sands are "an extremely carbon-intensive form of energy and a serious cause of local environmental pollution" (AXA 2018, 16). AXA credits its coal and oils sands divestment with brining its investment portfolio more in line with the Paris Agreement (AXA 2020, 19).

In addition to targeted coal and oils sands divestment efforts, there are also increased efforts to divest from or restrict investment in fossil fuel assets more generally. In 2019, citing "financial risk," the University of California system stated that it would no longer invest in fossil fuel assets. The classifications that the University of California will use to identify what constitutes a fossil fuel asset are not entirely clear (Mello 2019). Presumably, further fossil fuel restrictions could include refraining from investments in those assets that the University of California has already divested from, other types of fossil fuels, or expanded definitions of thermal coal and oil sands. The University of California previously divested from both coal and oil sands (Mello 2019).

While companies themselves as asset managers continue to divest from fossil fuel assets, some government entities are also asserting that they will not hire companies invested in or doing business with certain fossil fuel industries. For instance, in 2018, the Paris City Council passed a motion asking insurers to stop any support for the coal industry (Insurance Business 2018; Resolution Urging Coal Divestment 2018; Paris City Council Motion 2018). Later that year, the City and County of San Francisco took the position that it would also no longer do business with insurers that are invested in specific fossil fuel–related assets, in particular coal and oil sands, or underwriting their operations. The City and County of San Francisco Board of Supervisors passed a resolution urging the City and County of San Francisco to screen potential insurers for thermal coal and oil sands investments and to end any formal relationship with insurers that are not taking steps toward full divestment from thermal coal and oil sands (Resolution Urging Coal Divestment 2018). The classifications that the City and County of San Francisco Board of Supervisors will use to identify thermal coal and oil sands have not currently been released publicly.

4.2.5 Comparing classifications

With continued efforts to monitor assets that could potentially become stranded, there have been increasing divergences in what constitutes fossil fuel assets. Some of these divergences are observable, for instance, on the face of the classifications for the three efforts that were previously discussed: the California pension fund classification, the Climate Risk Carbon Initiative classifications, and the Global Coal Exit List classifications. The California pension fund classification, the Climate Risk Carbon Initiative classifications, and the Global Coal Exit List classifications all assert they are tracking coal. However, each effort defines what constitutes a coal asset slightly differently.

Both the California pension fund and Climate Risk Carbon Initiative classifications specifically apply to thermal coal and exclude metallurgic coal. In principle, the Global Coal Exit List follows the same approach. However, if companies do not differentiate between revenue generated from thermal coal and metallurgic coal, the Global Coal Exit List calculates the total coal share of revenue. While a classification system including all coal is presumably more inclusive, it is unclear how many companies would fit into a classification system tracking all coal and not one tracking simply thermal coal. That may also depend on what business activities the classification system tracks and what threshold percentage is applied.

The California pension fund classification specifically states that it applies to mining operations. The Climate Risk Carbon Initiative and the Global Coal Exit List do not just track mining operations. They also track other upstream and downstream activities. Other upstream activities include coal exploration and the development of coal mines. Downstream activities include the transportation, trading, and burning of coal.

The California pension fund classification includes companies generating at least fifty percent of their revenue from thermal coal. The Climate Risk Carbon Initiative and the Global Coal Exit List instead look at where revenue is at least thirty percent. The latter presumably includes a more extensive list of companies. Also, the Climate Risk Carbon Initiative and the Global Coal Exit List track all coal-related business activities. There may, however, be variation between the lists with regard to how much business activity is being generating from these lines of activities, for instance, where a mixed mining operation is mining different types of metals, an equipment manufacturer is providing equipment to different types of business activities, or a transportation company is moving various goods. Such distinctions may depend on available data and could change over time.

The Global Coal Exit List tracks mining operations that produce large quantities of coal. For instance, this could include large international companies with various lines of business, one of which is significant coal mining, even where coal is not responsible for thirty percent or more of revenue. The California pension fund and the Climate Risk Carbon Initiative classifications do not specifically track this.

Both the Climate Risk Carbon Initiative and Global Coal Exit List classifications include utilities that generate thirty percent or more of their electricity from the burning of thermal coal. The California pension fund classification does not track utilities, however. The California pension fund classification only looks at the mining of thermal coal.

The Global Coal Exit List tracks utilities that generate a large amount of electricity from coal. For instance, this could include large utilities that do not generate nearly thirty percent of their electricity from coal because they rely on a diversified energy mix, but they nonetheless still generate a significant aggregate amount of electricity from the burning of thermal coal. The California pension fund and the Climate Risk Carbon Initiative classifications do not specifically track this.

The Global Coal Exit List tracks utilities that are building new coal power plant facilities. The California pension fund and the Climate Risk Carbon Initiative classifications do not track this. It is unclear to what extent this classification draws in additional companies. It may depend on the global reach of covered investments, because the limited coal power development taking place is primarily located in Asia.

There are also differences among efforts to track other types of fossil fuel assets. For instance, the Climate Risk Carbon Initiative uses a threshold approach for tracking other fossil fuel assets that is similar to how the

Initiative tracks thermal coal assets, with revenue and power generation thresholds. However, other efforts that are still developing are specifically targeting oil sands. The classifications for determining what companies constitute oil sands companies are still unclear, but there will likely be divergences, as has been seen with coal, if targeting this asset becomes more widely accepted.

As discussed previously with respect to coal, aggregate classifications for enterprises and utilities are not always used in systems for monitoring assets that could potentially become stranded. They are also not necessarily explained as incorporated, because they are viewed as identifying at-risk investments. Quantification systems exist for oil and gas but are not as widely adopted as those for coal. When oil and gas are included, they are sometimes a subset of assets such as oil sands. The California Department of Insurance, for instance, applied a threshold classification to oil and gas. Reinsurers and the City and County of San Francisco have specifically identified oil sands.

There seems to be the most agreement in classifying most upstream fossil fuel enterprises, such as extraction, production, and transportation of fossil fuels, as brown assets. However, even within this area, there are contentions about specific companies. Some classifications, such as those used by California pensions, are narrow. They only include companies' mining operations or those generating over fifty percent of their revenue from thermal coal.

When applying only threshold classifications, it can also be difficult to classify companies that are involved in a range of activities. This includes certain companies transporting various goods or manufacturing equipment for a range of activities, not just fossil fuel enterprises. Also, categorizing fossil fuel enterprises as a specific type, such as thermal coal, can also be difficult when there are companies that operate mixed mining facilities or the raw materials could be used for separate activities, such as the production of steel rather than the generation of electricity.

There is disagreement in classifying utilities at fossil fuel companies that extends beyond the distinctions discussed previously. The two most established databases, California's Climate Risk Carbon Initiative database and the Global Coal Exit List, both recognize utilities as fossil fuel companies, although both identify that as one particular classification among several for identifying fossil fuel assets. Alternatively, some rating agencies include energy mix as one particular classification among several in categorizing a utility as at risk. However, they also view other factors, such as market position and local regulator policies, as important in understanding at-risk conditions. There is the thinking that some facilities have the ability to transfer away from the burning of coal, so in one year, they could be labeled thermal coal, and that labeling would not necessarily be appropriate in the following year. For example, Moody's has incorporated energy mix as part of the rating criteria for regulated electric and gas utilities. In particular, the rating looks at generation and fuel diversity. Utilities with lower exposure to penalties and taxes typically connected with the fossil fuel industry are typically rated higher. However, the generation and fuel diversity factor only composes five percent of a utility's rating (Moody's 2017, 33). A number of other factors are weighted more heavily (Moody's 2017, 4).

With efforts to monitor fossil fuel assets, there are additional challenges that present themselves. To begin with, the organizational structure of a company is not always clear. This can become an issue when applying threshold classifications that may not identify a parent company as a fossil fuel asset but instead identify a subsidiary company. To address this concern, the California Department of Insurance issued specific guidance indicating that its efforts applied to all companies, regardless of their status as a parent or subsidiary (CDI 2016, 5). The Global Coal Exit List also specifically states that it includes both parent and subsidiary companies, listing information for both (Urgewald n.d.a).

It is difficult to identify the investment holdings of specific companies from publicly available information. In many countries, companies are not required to publicly release their investment holdings. Even in those countries where certain companies are required to publicly release this information, such as the United States, it is still not easy to identify all investment holdings. For instance, it is not easy to identify private placement information that is not publicly traded with corresponding security identification numbers. At a minimum, private placement information needs to be culled manually from disclosures and cannot simply be uploaded along with information for publicly traded investment holdings.

Threshold information is proprietary. With respect to revenue, companies may take issue with how they are being classified, but it would be difficult for them to identify a concern, because it may not be clear to them how they are being categorized and how such a distinction is being made. With respect to power generation, there is not even a readily available public list of how all utilities are generating their electricity.

Tracking divestment of specific types of assets is extremely difficult. It is not difficult to observe the total aggregate amount. However, aggregate amounts can fluctuate based on the price of the assets rather than the amount of securities being held. Moreover, a company could report divesting from one fossil fuel security only to purchase another fossil fuel security.

In addition to how monitoring is conducted, there are multiple means to address concerns with assets at risk of becoming stranded assets. While some companies believe that divestment is the best means for addressing concerns with at-risk assets, other entities believe that, at least in certain circumstances, it is more appropriate to engage companies rather than simply divest from them. This make sense, for instance, in circumstances where a current stock position makes it difficult to currently divest without suffering significant losses or where a company investee is positioned in such a manner that it could modify its business operations to address the factors that make the company an at-risk asset. While smaller investors may be unable to exert necessary influence on a company investee, a larger investor may be able to do so.

As two of the world's largest pension funds and asset holders, CalPERS and CalSTRS have significant influence on impacting market behaviors. As seen simply with its coal effort, CalPERS reached out to seventeen companies asking that they divest from thermal coal if the company wished for CalPERS to remain invested in it. While it is unclear to what extent these outreach efforts impacted the investee companies' decision making, three of the companies that CalPERS engaged provided a plan to transition away from thermal coal (CalPERS 2017, 3). CalPERS is also now leading a larger corporate engagement strategy known as Climate Action 100+ where it and other large asset holders are asking investee companies to address climate change on an even more robust scale.

The NGFS explained that it is important to have clear standards for monitoring investments at risk of becoming stranded, such as investments in fossil fuel companies. Such standards, which the NGFS hinted at as a "brown" taxonomy, could help identify the assets that will be impacted by the Paris Agreement and a transition to a low-carbon economy (NGFS 2019, 35). A recent publication by the Bank of International Settlements and Bank of France also discussed the idea of a brown penalizing factor that would increase capital requirements where there is greater exposure to investments at risk of becoming stranded (Bolton et al. 2020, 52). At this time, however, financial regulators have not collectively endorsed such a taxonomy or a penalizing factor.

References

Aravindan, Aradhana, and John Geddie. 2020. "Singapore Aims to Phase Out Petrol and Diesel Vehicles by 2040." *Reuters*, February 18. www.reuters.com/article/us-singapore-economy-budget-autos/singapore-aims-to-phase-out-petrol-and-diesel-vehicles-by-2040-idUSKBN20C15D.

AXA. 2018. *Climate-Related Investment & Insurance Report*. Paris: AXA.

AXA. 2020. *2020 Climate Report: Renewed Action in a Time of Crisis*. Paris: AXA.

Bolton, Patrick, Morgan Despres, Luiz Awazu Pereira da Silva, Frédéric Samama, and Romain Svartzman. 2020. *The Green Swan: Central Banking and Financial Stability in the Age of Climate Change*. Basel: Bank of International Settlements and Bank of France.

British Broadcasting Corporation. 2017. "China Looks at Plans to Ban Petrol and Diesel Cars." *British Broadcasting Corporation*, September 10. www.bbc.com/news/business-41218243.

British Broadcasting Corporation. 2020. "Germany Agrees Plan to Phase Out Coal Power by 2038." *British Broadcasting Corporation*, January 16. www.bbc.com/news/world-europe-51133534#:~:text=The%20German%20government%20and%20regional,depending%20on%20the%20progress%20made.

Brody, Sarah, Matt Rogers, and Giulia Siccardo. 2019. "Why, And How, Utilities Should Start to Manage Climate-Change Risk." *McKinsey & Company Electric Power & Natural Gas*, April. www.mckinsey.com/industries/electric-power-and-natural-gas/our-insights/why-and-how-utilities-should-start-to-manage-climate-change-risk#.

Buckley, Tim. 2019. *Over 100 Global Financial Institutions Are Exiting Coal, With More to Come*. Cleveland: Institute for Energy Economics and Financial Analysis.

California Department of Insurance. 2016. *Climate Risk Carbon Initiative Questions & Answers*. Sacramento: California Department of Insurance.

California Department of Insurance. 2018. *Survey Results of Insurer Investments Tied to Thermal Coal Released*. Sacramento: California Department of Insurance.

California Department of Insurance. n.d. "Climate Risk Carbon Initiative." Accessed July 25, 2020. https://interactive.web.insurance.ca.gov/apex_extprd/f?p=250:1:0.

California Public Employees' Retirement System. 2017. *Public Divestiture of Thermal Coal Companies Act Report to the California Legislature and Governor*. Sacramento: California Public Employees' Retirement System.

California Public Employees' Retirement System. n.d. "Investments." Accessed July 25, 2020. www.calpers.ca.gov/page/investments.

California State Teachers' Retirement System. 2017. "CalSTRS Takes Action to Divest of All Non-U.S. Thermal Coal Holdings." *Modified*, June 7. www.calstrs.com/news-release/calstrs-takes-action-divest-all-non-us-thermal-coal-holdings.

California State Teachers' Retirement System. n.d. "Investments Overview." Accessed July 25, 2020. www.calstrs.com/investments-overview.

Carbon Tracker. 2011. *Unburnable Carbon – Are the World's Financial Markets Carrying a Carbon Bubble?* London: Carbon Tracker.

Carbon Tracker. 2019. *Balancing the Budget: Why Deflating the Carbon Bubble Requires Oil & Gas Companies to Shrink*. London: Carbon Tracker.

Carney, Mark. 2015. "Breaking the Tragedy of the Horizon – Climate Change and Financial Stability." Speech at Lloyd's of London, London, September.

Carney, Mark, Francois Villeroy de Galhau, and Frank Elderson. 2019. *Open Letter on Climate-Related Financial Risks*. London: Bank of England.

CUSIP Global Services. n.d. "CUSIP Global Services." Accessed July 25, 2020. www.cusip.com.

Examiner Staff. 2014. "SF State Joins Campaign to Divest Money from Fossil Fuels." *San Francisco Examiner*, April 8. www.sfexaminer.com/news/sf-state-joins-campaign-to-divest-money-from-fossil-fuels/?oid=2757805.

Gladstone, Rick. 2020. "Oil Collapse and Covid-19 Create Toxic Geopolitical Stew." *The New York Times*, April 22. www.nytimes.com/2020/04/22/world/middleeast/oil-price-collapse-coronavirus.html.

Government of the Netherlands. 2019. *Climate Agreement*. The Hague: Government of the Netherlands.

Harding, Robin. 2020. "Japan Vows to Slash Financing of Coal Power in Developing World." *Financial Times*, July 12. www.ft.com/content/482fa9e4-5eb5-4c61-a777-998993febae0.

Hill, Alice C., and Leonardo Martinez-Diaz. 2019. *Building a Resilient Tomorrow, How to Prepare for the Coming Climate Disruption*. New York: Oxford University Press.

Insurance Business. 2018. "Paris City Council Calls on Insurers to Ditch Coal." *Insurance Business*, May 3. www.insurancebusinessmag.com/uk/news/breaking-news/paris-city-council-calls-on-insurers-to-ditch-coal-99578.aspx.

Intergovernmental Panel on Climate Change. 2014. *Climate Change 2014: Synthesis Report*. Geneva: Intergovernmental Panel on Climate Change.

ISIN Organization. n.d. "ISIN Organization." Accessed July 25, 2020. www.isin.org.

Johnson, Boris. 2020. "PM Speech at COP 26 Launch." Speech at Prime Minister's Office, London, February.

Jones, Dave. 2016. *Coal Divestment & Carbon-Based Investment Data Call*. Sacramento: California Department of Insurance.

Krauss, Clifford. 2016. "Coal Production Plummets to Lowest Level in 35 Years." *The New York Times*, June 10. www.nytimes.com/2016/06/11/business/energy-environment/coal-production-decline.html.

Kuykendall, Taylor. 2015. "Roster of United States Coal Companies Turning to Bankruptcy Continues to Swell." *SNL Energy*, June 4. www.snl.com/InteractiveX/Article.aspx?cdid=A-32872208-12845.

Levisohn, Ben. 2017. "Why Paris Pullout Is No Cure All for Coal." *Barron's*, June 6. www.barrons.com/articles/why-paris-pullout-is-no-cure-all-for-coal-1496762504.

McCarran-Ferguson Act of 1945, 15 U.S.C. §§ 1011–1015 (1945).

Mello, Felicia. 2019. "Citing 'Financial Risk,' University of California Pledges to Divest from Fossil Fuels." *Cal Matters*, September 17. https://calmatters.org/education/higher-education/2019/09/uc-divests-fossil-fuels-citing-finance-renewable-energy-climate-change/.

Moody's Investors Service. 2017. *Regulated Electric and Gas Utilities*. New York: Moody's.

MSCI. n.d. "The Global Industry Classification Standard." Accessed July 25, 2020. www.msci.com/gics.

Network for Greening the Financial System. 2019. *A Call for Action, Climate Change as a Source of Financial Risk*. Paris: Network for Greening the Financial System.

Newsom, Gavin. Executive Order N-79-20 (2020).

Paris City Council Motion. 2018.

Petroff, Alanna. 2017. "These Countries Want to Ditch Gas and Diesel Cars." *CNN Business*, July 26. https://money.cnn.com/2017/07/26/autos/countries-that-are-banning-gas-cars-for-electric/.

Plumer, Brad. 2020. "In a First, Renewable Energy Is Poised to Eclipse Coal in U.S." *The New York Times*, May 13. www.nytimes.com/2020/05/13/climate/coronavirus-coal-electricity-renewables.html.

Public Retirement Systems: Public Divestiture of Thermal Coal Companies, S.B.185 (2015).

Regelink, Martjin, Henk Jan Reinders, Maarten Vleeschhouwer, and Iris van de Wiel. 2017. *Waterproof? An Exploration of Climate-related Risks for the Dutch Financial Sector*. Amsterdam: Dutch Bank.

Resolution Urging Insurance Industry Companies to Divest from Coal and Tar Sands Industries; And to End the Underwriting of Activities in Furtherance of the Extraction or Use of Coal and Tar Sands, Resolution No. 273–18 (2018).

Reuters. 2019. "France's EDF to Close Le Havre Coal-Fired Power Plant in Spring 2021." *Reuters*, June 7. www.reuters.com/article/france-electricity-coal/frances-edf-to-close-le-havre-coal-fired-power-plant-in-spring-2021-idUSL8N23E254.

Savage, Karen. 2019. "Shareholders Sue Exxon for Misrepresenting Climate Risks." *Climate Liability News*, August 7. www.climateliabilitynews.org/2019/08/07/exxon-shareholders-climate-risks-lawsuit/.

S&P Global. n.d. "Dow Jones U.S. Coal Index." Accessed 2020. www.spglobal.com/spdji/en/indices/equity/dow-jones-us-coal-index/#overview.

Urgewald. n.d.a. "Global Coal Exit List." Accessed July 25, 2020. https://coalexit.org.

Urgewald. n.d.b. "Urgewald." Accessed July 25, 2020. https://urgewald.org/english.

Yanguas, Paolo A., Gaurav Ganti, Robert Brecha, Bill Hare, Michiel Schaeffer, and Ursula Fuentes. 2019. *Global and Regional Coal Phase-Out Requirements of the Paris Agreement: Insights from the IPCC Special Report on 1.5°C*. Berlin: Climate Analytics.

5 Green taxonomies

Green taxonomies can help investors identify current green finance opportunities by classifying sustainable activities in a systematic manner.

There have been significant efforts to ensure the proper classification of sustainable investment opportunities and in some cases to encourage investment in these opportunities. Similar to how investments at risk of becoming stranded are identified, unique security identification numbers and third-party data products can be employed to determine whether particular securities are in a sector that is traditionally viewed as sustainable, such as renewable energy or electric vehicle production. Numerical identifiers and third-party data products can also be used to determine the percentage of revenue that companies are generating from enterprises that are traditionally viewed as within sustainable sectors and the percentage of electricity that utilities are generating from sources that are traditionally viewed as within sustainable sectors. Together the numerical identifiers and third-party data products can be used to run an automated analysis across an investment portfolio to quickly identify the prevalence of investments that have traditionally been viewed as sustainable.

However, because of concerns raised with the manner in which securities have traditionally been labeled sustainable, governments and nonprofit organizations are instead developing more robust methodologies for identifying sustainable or green investments rather than relying on exclusively on existing numerical identifiers and third-party data productions. Through stakeholder engagement, government and nonprofit organizations have been developing their own classification systems for identifying sustainable investments, which are generally referred to as green taxonomies. These taxonomies have allowed for the inclusion of investments in a greater number of sectors than simply those traditionally viewed as within sustainable sectors and have taken into account additional considerations. These taxonomies are currently mapping to sector information, and third-party data product providers have been developing third-party data products that will further link green taxonomy classifications with security identification numbers. As these tools develop, numerical identifiers and new third-party data products can be used to run an automated analysis across an investment portfolio to quickly identify the prevalence of investments that are viewed as sustainable under green taxonomies.

Notable among efforts to classify green investments is the development and implementation of internationally recognized Green Bond Principles. There have also been efforts to harmonize green bond taxonomies currently in use, in particular between the Climate Bonds Initiative (CBI) and that of the Chinese government. A European Union Green Bond Standard is also being developed as part of its larger European Union sustainable finance strategy that includes a European Union taxonomy.

5.1 Chinese efforts

5.1.1 Guidance on Green Loans

The Chinese government has been the first government to aggressively push green investments. The Chinese government first encouraged green lending by developing green loan guidance. In 2013, the China Banking and Insurance Regulatory Commission established standards for green loans through its Guidance on Green Loans (CBIRC 2012; NGFS 2019, 34). After the publication of this Guidance, the China Banking and Insurance Regulatory Commission has required all large banks to report twice a year on the balance of and environmental benefits from a dozen categories of green loans. This Guidance has helped lead to a significant amount of green loan activity, with China's banks now having over a trillion dollars or approximately ten percent of their loan balance in green investments (CBIRC 2012; NGFS 2019, 34).

5.1.2 Green Bond Endorsed Project Catalogue

The Chinese government followed up its green loan guidance by introducing the first national-level green bond taxonomy. In 2015, the People's Bank of China's Green Finance Committee of the China Society for Finance and Banking published the Green Bond Endorsed Project Catalogue (PBOC 2015; NGFS 2019, 34). The Catalogue detailed six main categories and thirty-one subcategories of projects eligible for green bond financing (PBOC 2015; NGFS 2019, 34). The six main project categories are listed in Table 5.1.

Although use of the Catalogue is not mandatory, essentially all Chinese issuers, investors, and verifiers use it. In only a few years, Chinese financial institutions have issued over one hundred billion dollars in green bonds, accounting for roughly a quarter of green bonds issued from 2016 to 2018. This has made China one of the world's biggest green bond markets. Chinese regulators have also introduced rules and guidelines on green bond disclosures and verification (PBOC 2015; NGFS 2019, 15, 34).

Table 5.1 Chinese Green Bond Endorsed Project Catalogue – Green Classification

Energy Savings
Pollution Prevention and Control
Resource Conservation and Recycling
Clean Transport
Clean Energy
Ecological Protection and Climate Change

Even with the onset of the COVID-19 pandemic, China is still focused on ensuring a transition to a low-carbon economy. Led by General Secretary and President Xi Jinping, the Politburo Standing Committee has dedicated over a trillion dollars to a range of low-carbon technologies (Worland 2020). It is not entirely clear how funding will be connected with China's guidance on green loans and the Green Bond Endorsed Project Catalogue, but it is likely that capital will be directed to projects identified as green under these efforts.

5.2 Nonprofit efforts

5.2.1 Green Bond Principles

Outside of China, there have been a number of other efforts to label financial assets as environmentally sustainable or green, especially in private capital markets. This has helped direct investment to such assets, including green loans, green bonds, and green funds. With a number of organizations identifying assets as green and with little oversight as to what constituted green, there were growing concerns over organizations inappropriately labeling securities as sustainable or green, or "greenwashing." To address these concerns, several international investment banks collaborated in 2014 to create voluntary best practice guidelines for the asset type with the most developed taxonomies and markets, green bonds (EU 2019b, 16). The voluntary best practice guidelines, called the Green Bond Principles, better ensure proper certification of bonds as green, with funds being directed appropriately to specific projects. The Green Bond Principles detail best practices that include guidance on the use of proceeds, project evaluation and selection, proceed management, and reporting (GBP and ICMA 2018; CBI 2020).

To begin with, the Green Bond Principles specifically recognize several broad categories of eligible green projects that are appropriate for proceeds, including the following: climate change mitigation, climate change adaptation, natural resource conservation, biodiversity conservation, and pollution prevention and control (GBP and ICMA 2018, 4). These project categories are displayed in Table 5.2.

Table 5.2 Green Bond Principles – Green Classification

Climate Change Mitigation
Climate Change Adaptation
Natural Resource Conservation
Biodiversity Conservation
Pollution Prevention and Control

Next, the Green Bond Principles encourage transparency for project evaluation and selection. First, a green bond issuer should clearly explain the environmental sustainability objectives to investors. Second, a green bond issuer should explain to investors the process used to determine how a project fits within eligible green bond project categories. Third, a green bond issuer should identify any related eligibility criteria. The Green Bond Principles further recommend supplemental external review for these considerations (GBP and ICMA 2018, 5).

Further, the Green Bond Principles provide detailed guidance on the management of proceeds. This includes the issuer segregating net proceeds of a green bond to a subaccount, moving them to a subportfolio, and formally attesting to them. As long at the green bond is outstanding, the issuer should also periodically adjust the balance of the net proceeds to match allocations made to eligible green bond projects and update investors on the intended types of temporary placement for the balance of unallocated net proceeds. The Green Bond Principles also recommend that an auditor or other third party verify the internal tracking methodology and allocation of funds from the green bond proceeds (GBP and ICMA 2018, 5).

Finally, the Green Bond Principles discuss favored reporting processes. The Green Bond Principles support the use of annual reports showing a list of projects where green bond proceeds have been allocated, with a brief description of the projects, the amounts of proceeds allocated, and their expected impact. Until the full allocation of proceeds, the Green Bond Principles also dictate that an issuer should be able to readily provide current information on a timely basis in case of material developments. With respect to how information is categorized, the Green Bond Principles recommend the use of at least transparent qualitative performance criteria and transparent quantitative performance criteria where feasible. For certain project types, there are also voluntary guidelines aimed at facilitating more consistent reporting (GBP and ICMA 2018, 5).

The European Investment Bank, which is owned by the countries within the European Union, chairs the Steering Committee of the Green Bond Principles. The European Investment Bank is also the world's largest issuer of green bonds (EIB and GFC 2017, 20). An independent secretariat hosted by the International Capital Market Association monitors and develops guidelines for the Green Bond Principles (EU 2019a, 16, 32–35).

5.2.2 Climate Bonds Initiative

Certification programs label debt instruments as green bonds when they are issued and proceeds are maintained in conformance with green bond practices, such as those described within the Green Bond Principles. One of the most prominent certification programs is the CBI certification program. The CBI certification program essentially takes the broad structure articulated in the Green Bond Principles and creates a workable classification system for labeling green bonds. The CBI certification program uses a "Climate Bonds Certified" label to identify debt instruments that can appropriately be labeled as green bonds under the Green Bond Principles and any additional requirements imposed by the CBI (CBI 2017, 3). This label provides investors with confidence that they are purchasing a green asset when buying such a debt instrument.

In order for a debt instrument to obtain a Climate Bonds Certified label, an issuer needs to have a third-party CBI approved verifier provide a verified statement that a bond meets the CBI standard (CBI 2017, 4–5). To meet the CBI standard, the proceeds from a debt instrument must fund a CBI eligible project. This includes a project that conforms with a CBI taxonomy that is essentially a detailed sector and subsector project listing that corresponds with requirements under the Green Bond Principles. This includes eight categories of projects: energy, transport, water, buildings, land use and marine resources, industry, waste, and information and communications technology. There are also subcategories to each of these with their own corresponding guidance (CBI 2017, 8, 15). In addition, to meet the CBI standard, use of proceeds, tracking, and reporting need to abide by CBI mandatory requirements that are modeled on the Green Bond Principles (CBI 2017, 3–13).

In a TCFD and French Energy Transition Law aligned report, AXA explained how the CBI label is used in practice. AXA has invested several billion dollars in green bonds making sure that these bonds are externally labeled by CBI and rating agencies that confirm that CBI taxonomy has been appropriately used and the proceeds are properly used. AXA also makes sure that bonds satisfy additional criteria that are tailored for specific project types (AXA 2020, 31).

The NGFS has explained ideals for green taxonomies. According to the NGFS, policymakers should ensure that a taxonomy is robust and detailed enough to allow for the certification of green assets and investments projects. A taxonomy should also prevent greenwashing (NGFS 2019).

Policymakers are working on harmonizing taxonomies in different jurisdictions. For instance, in 2017, the China Green Finance Committee and the European Investment Bank published a joint white paper entitled "The Need for a Common Language in Green Finance." This was followed by a second edition in 2018. The white paper effort compares different green bond taxonomies and describes a potential strategy for harmonization of green bond taxonomies. This includes the work of the European Investment Bank to create translations for taxonomies so that categories of projects are comparable (EIB and GFC 2017).

5.3 European Union efforts

The European Commission has also been developing a sustainable investment strategy (EC 2018). This sustainable investment strategy is part of the European Commission's long-term sustainability objective of achieving carbon neutrality by 2050. The European Commission believes that the European Union needs to fill an investment gap of nearly three hundred billion dollars per year in order to reach its long-term sustainability objective of achieving carbon neutrality, and the European Commission's sustainable investment strategy is necessary to achieve this objective (EU 2019b, 85, 2020a, 8).

5.3.1 European Union taxonomy

The European Commission established a Technical Expert Group on sustainable finance to assist with developing legislative proposals for its sustainable finance strategy. This Technical Expert Group includes thirty-five members from fields such as academia and finance and observers from the European Commission and international bodies. The legislative proposals include a European Union taxonomy on when it is appropriate to classify financial activities as sustainable or green and a European Union green bond standard (EU 2019c, 16–18, 2020a, 10–12).

The European Union taxonomy has an overarching classification system for determining whether a financial activity can be identified as environmentally sustainable. For an economic activity to be identified as environmentally sustainable, it needs to (i) substantially contribute to at least one of six environmental objectives, (ii) do no significant harm to any of the other six environmental objectives, and (iii) comply with minimum safeguards (EU 2019c, 19, 2020a, 2). The six environmental objectives are climate change mitigation, climate change adaptation, sustainable use and protection of water and marine resources, transition to a circular economy, pollution prevention and control, and protection and restoration of biodiversity and ecosystems (EU 2019c, 19, 2020a, 2). To comply with minimum safeguards, financial activity must also abide by the UN Guiding Principles on Business and Human Rights and the OECD Guidelines for Multinational Enterprises, including the principles and rights set out in the eight fundamental conventions identified in the International Labor Organization's declaration of Fundamental Rights and Principles at Work and the International Bill of Human Rights (EU 2019c, 19, 2020a, 2).

The plan for the European Union taxonomy and related technical screening criteria is what will act like a dictionary for sustainable activities (EU 2019c, 19, 2020a, 51–55). A user of the taxonomy can look up a particular sector or subsector and determine the criteria and thresholds necessary to meet in order to show meeting one of the six environmental objectives that has been identified by the European Commission as important for achieving its long-term sustainable objective. A user of the taxonomy can also look up a particular sector or subsector and determine the criteria to find guidance that can be used to help ensure that the other environmental objectives are not significantly harmed in meeting an environment objective. If an activity is not classified, it does not mean that it is brown or not sustainable, it just means that the activity has not currently been classified. For a fund, rather than a single activity, a user would be able to identify the percentage of the fund that fits the European Union taxonomy.

The European Union taxonomy and especially related technical screening criteria are dynamic. They will develop over time through consultation with stakeholders. They will require continuous updates to be kept current, taking into account new policy and technological innovations (EU 2019c, 22, 2020a, 54–55). A recently released Taxonomy Technical Report was several hundred pages in length, with the majority of pages dedicated to information on technical screening criteria for specific sectors and subsectors. This detailed how the taxonomy could be applied to sectors and subsectors and provided corresponding industry classifications (EU 2019c, 107–414, 2020b).

Financial institutions are not required to use the European Union taxonomy, but it must be used when marketing financial products as environmentally sustainable or green. Along these lines, European Union countries that are setting up policies for the marking of financial products as financially sustainable or green must also take into account the European Union taxonomy when doing so (EU 2019c, 57–59, 2020a, 26). In the future, investors may also put pressure on financial institutions to disclose their activities according to the European Union taxonomy.

5.3.2 European Union Green Bond Standard

As the centerpiece of the European Commission's sustainable finance strategy, the European Union taxonomy sets up a classification system for another important part of the European Commission's sustainable finance strategy, the European Union Green Bond Standard. In essence, the European Union has developed a European Union Green Bond Standard that allows for certification of green bonds in connection with the guidance of the European Union taxonomy. The Technical Expert Group recommends that this European Green Bond Standard should be a standard that most green bond issuers will be able to comply with for new issuances and which could even allow issuers the option of aligning their existing bonds (EU 2019a, 16–17, 2019b, 23–24, 2020c, 11).

As far as the structure of the green bond market itself, the Technical Expert Group has recognized that the European Union Green Bond Standard should be built on market best practices (EU 2019b, 23, 2020c, 11). Today, the green bond market consists of over six hundred billion dollars in capital, with major issuers based in Europe, North America, and Asia-Pacific (EU 2019b, 16, 2020c, 11). The creation of the Green Bond Principles and their adoption by the vast majority of issuers greatly facilitated standardization within the green bond market (EU 2019b, 16). The Technical Expert Group acknowledged that European and international green bond markets generally operate appropriately through best practices embodied within the Green Bond Principles. Therefore, while the structure for evaluating eligible projects differs, the Green Bond Principles are the starting point for the European Union Green Bond Standard with similar procedures for issuing, verifying, and reporting on green bonds (EU 2019b, 23–24, 57–61). Relying as much as possible on an existing green bond taxonomy and best practices in the green bond market allows for the European Union Green Bond Standard to be actionable and operational in the short term (EU 2019b, 23–24).

With respect to the classification system for eligible projects, green bond proceeds must be allocated only to green projects, which are defined as contributing substantially at least one of the European Union taxonomy's six environmental objectives while not significantly harming any of the European Union's other environmental objectives and complying with the minimum safeguards represented in the UN Guiding Principles on Business and Human Rights and the OECD Guidelines for Multinational Enterprises, including the principles and rights set out in the eight fundamental conventions identified in the International Labor Organization's declaration of Fundamental Rights and Principles at Work and the International Bill of Human Rights (EU 2019b, 57–58, 2020c, 14). Eligible green projects are also expected to be aligned with technical criteria issued for the European Union taxonomy (EU 2019b, 58, 2020c, 14). Specifically, where technical screening criteria have been developed for specific sectors and subsectors, eligible green projects shall align with these criteria, except in exceptional circumstances where the criteria may not be directly applicable because of factors such as a project's innovative nature, complexity, or location (EU 2019b, 58, 2020c, 14). It is further understood that subsequent changes to the European Union taxonomy will not apply to green bonds that have already been issued. Their status as green bonds will effectively be grandfathered (EU 2019b, 59, 2020c, 32). In Table 5.3 are the three primary steps for green bond classification that model the European Union taxonomy classification system.

Table 5.3 European Union Green Bond Standard – Green Classification

STEP 1
Contributes Substantially to at Least One European Union Environmental Objective
Climate Change Mitigation
Climate Change Adaptation
Sustainable Use and Protection of Water and Marine Resources
Transition to a Circular Economy
Pollution Prevention and Control
Protection and Restoration of Biodiversity and Ecosystems
STEP 2
Not Significantly Harm Any Other European Union Environmental Objective
Climate Change Mitigation
Climate Change Adaptation
Sustainable Use and Protection of Water and Marine Resources
Transition to a Circular Economy
Pollution Prevention and Control
Protection and Restoration of Biodiversity and Ecosystems
STEP 3
Comply With the Minimum Safeguards
An asset must also comply with the UN Guiding Principles on Business and Human Rights and the OECD Guidelines for Multinational Enterprises, including the principles and rights set out in the eight fundamental conventions identified in the International Labor Organization's declaration of Fundamental Rights and Principles at Work and the International Bill of Human Rights

A financial institution marketing a financial product as green or sustainable is responsible for ensuring that all three steps are met and that the financial product is not at odds with published European Union taxonomy technical screening criteria unless otherwise justified. Presumably, there will be areas of contention over whether certain steps are satisfied. Contentions are possibly likely to occur where there is a conflict between the first and second or third steps, especially where the European Union taxonomy technical screening criteria does not provide specific guidance. As time passes, however, it is likely that European Union taxonomy technical screening criteria will work to resolve some of these contentions. Moreover, third-party data product providers will develop tools linking security identification numbers with green classifications. For equity, this would likely consist of a percentage of greenness. For debt, this would likely consist of either a green classification or no green classification.

Issuing green bonds should occur in the following manner. An issuer of green bonds shall produce an issuing framework. The framework shall describe, for instance, how its issuing strategy aligns with the objectives of the European Union Green Bond Standard and the most current version of the European Union taxonomy, the green projects to be financed or refinanced with green bonds, and reporting detailing the methods for doing so and the frequency. The issuer should publish its framework before or when issuing green bonds and make the report available until the maturity of the green bonds (EU 2019b, 28–31, 2020c, 20–24).

Verifying green bonds should occur in the following manner. Issuers should appoint external reviewers to confirm alignment with the European Union Green Bond Standard and the most current version of the European Union taxonomy. This shall take place at the time green bonds are issued and after the allocation of green bond proceeds. Issuers shall make external reviewer reports publicly available. External reviewers will be subject to accreditation with regard to standardize procedures, minimum qualifications, and professional codes of conduct (EU 2019b, 31–42, 2020c, 26–28).

Reporting for green bonds should occur in the following manner. Issuers shall report at least annually until bond proceeds have been fully allocated to green projects and thereafter if there is a material change in the allocation of proceeds. The contents of a report shall include items such the following: a statement of compliance with the European Union Green Bond Standard, the amount allocated to each green project category, the geographic distribution of green projects, the actual or estimated environmental impact of green projects, what share of proceeds is for financing and for refinancing, and the total amount of green bonds outstanding divided by the total amount of debt outstanding at the end of a reporting period (EU 2019b, 28–31, 2020c, 24–26).

The Technical Expert Group has recommended that the European Union Green Bond Standard begin largely as a voluntary effort. To achieve this aim, the Technical Expert Group is advising that the European Union encourage the adoption of the European Union Green Bond Standard and require corresponding reporting. The European Union should specifically encourage institutional investors to adopt the European Union Green Bond Standard requirements when designing their green fixed-income investment plans and communicate their plans to green bond issuers and underwriters. The European Union itself should adopt a disclosure structure for institutional investors that requires institutional investors to periodically disclose green bond holdings. The European Union should also encourage underwriters to disclose what portion of underwritten bonds are green bonds (EU 2019b, 43–51, 2020c, 31, 44–46).

According to the Technical Expert Group, the European Union should also encourage efforts that would subsidize costs and otherwise stimulate market participation (EU 2019b, 43–51). The European Central Bank should consider prioritizing investment in green bonds certified under the European Union Green Bond Standard when making green bond purchases, so long as doing so does not prejudice the principle of market neutrality (EU 2019b, 45–46). The Technical Expert Group is advising that the European Union encourage all public-sector bond issuers in countries within the European Union, regardless of whether they have previously issued green bonds, to use the European Union Green Bond Standard for future green bond issuances (EU 2019b, 47–48). The European Commission could also consider credit enhancements for green bonds certified under the European Union Green Bond Standard, establish a grant program to pay the external verification costs for issuers, foster the growth of accredited verifiers, and promote the development of best practices within and outside the European Union (EU 2019a, 29–36, 2019b, 43–51).

The European Union taxonomy and European Union Green Bond Standard will likely have far-reaching impacts both within and outside the European Union. The European Union Green Bond Standard sets up a harmonized view across the countries of the European Union. Outside of the European Union, other green bond taxonomies are modeling to the European Union Green Bond Standard. For instance, the CBI has announced plans to harmonize its taxonomy with the European Green Bond Standard. Both efforts already have many similarities in their operation, as they are modeled on the Green Bond Principles, so presumably efforts at harmonizing will largely be centered on ensuring that the CBI taxonomy of eligible projects includes projects that are eligible under the European Union Green Bond Standard.

Even with the onset of the COVID-19 pandemic, the European Union is increasingly focused on ensuring a transition to a low-carbon economy. German Chancellor Angela Merkel emphasized, "We also have to encourage each other not to forget climate protection. . . . On the contrary, enhance it" (Worland 2020). Along these lines, Germany is evaluating major European companies again the European Union taxonomy to facilitate the develop of "sustainable, climate-friendly" recovery programs for a post–COVID-19 European Union (Azizuddin 2020; Federal Ministry for the Environment, Nature Conservation and Nuclear Safety 2020). The European Commission is dedicating over a trillion dollars to a European Green Deal that European Commission President Ursula von der Leyen has referred to as the European Union's "motor for the recovery" (Worland 2020).

5.4 Comparing green classifications

Although not necessarily all projects would be eligible under each classification system, the three primary classification systems for evaluating green bonds have very similar overarching categories for evaluating eligible green projects. Table 5.4 compares categories of eligible green projects. The major difference with the classification systems is that the European Union Green

Table 5.4 Comparison of Green Classification Green Project Categories

	Chinese Green Bond Endorsed Project Catalogue	*Green Bond Principles*	*European Union Green Bond Standard*
Climate Change Mitigation	Energy Savings, Clean Transport, and Clean Energy Each Separate Project Categories	A Project Category	A Project Category
Climate Change Adaptation	Not a Project Category	A Project Category	A Project Category
Waste Reduction and Recycling	A Project Category	Not a Project Category	A Project Category
Water Conservation	Not a Project Category	Not a Project Category	A Project Category
Pollution Prevention	A Project Category	A Project Category	A Project Category
Healthy Ecosystems	A Project Category	A Project Category with Biodiversity Identified on Its Own	A Project Category with Biodiversity

Bond Standard has a couple of additional requirements. It does not allow for projects that harm another green objective, and projects must comply with minimum safeguards.

5.5 Emerging efforts

Other national and subnational governments have also been referencing green taxonomies used elsewhere in the world or are developing their own green taxonomies. The United House of Representatives introduced House Resolution 109 titled "Recognizing the duty of the Federal Government to create a Green New Deal." The resolution acknowledged a need to keep temperatures below 1.5 degrees Celsius above preindustrialized levels in order to avoid the most severe climate impacts and contemplated efforts to make major reductions in greenhouse gas emissions. Although not specifically mentioning a green taxonomy, contemplated efforts for reducing greenhouse gas emissions include "directing investments to spur economic development" (Green New Deal 2019).

California also announced that it would investigate a framework for green investments. Governor Gavin Newsom directed the California Department of Finance to investigate potential green investments for CalPERS and CalSTRS (Newsom 2019). More details on how green investments will be categorized should become public over time.

Japan has been working on green bond guidelines and investing in green bonds. Japan's Ministry of the Environment published original Green Bond Guidelines in 2017 (MoE 2017) and published updated Green Bond Guidelines in 2020 (MoE 2020). Japan's 1.5 trillion-dollar pension fund announced that it is planning to make substantial green investments. The pension fund has been planning to create a strategy for making these investments (Mishima 2019). Additional details will be decided as these efforts develop.

The Monetary Authority of Singapore established a green investment program. The program appears to contemplate the development of environmental risk and management guidelines and the development of plans to involve external reviewers and rating agencies (MAS 2019). This program seeks to invest two billion dollars in green investments. Additional details will be decided as these efforts develop.

Moreover, while there has not been a lot of enforcement against those inappropriately asserting that their financial instruments are green, government agencies are starting to take steps to ensure that bond issuers are properly identifying debt as green. The Financial Conduct Authority recently explained that it would not tolerate greenwashing and will instead "take appropriate action to prevent consumers [from] being misled" (Binham and Hook 2019). Ben Caldecott, who directs Oxford University's sustainable finance program, explained the significant of this described action: "This is a really big deal, because firms are making all sorts of claims. Consumers need to be protected. The [Financial Conduct Authority] needs to make sure these claims are justifiable and it is beginning to set out a framework for how it will do that" (Binham and Hook 2019).

Just as there are discussions about a brown penalizing factor, there are also discussions about a green supporting factor. However, while the Chinese government has set up requirements to make green investments, other bank supervisors and regulators have expressed reluctance to take such an approach. A recent publication by the Bank of International Settlements and Bank of France stated that "it is not obvious why being exposed to 'green' sectors would necessarily reduce non-climate-related financial risks, and thereby justify lower capital requirements" (Bolton et al. 2020, 52). Green investments are still subject to transition and physical risks (Bolton et al. 2020, 52).

References

AXA. 2020. *2020 Climate Report: Renewed Action in a Time of Crisis*. Paris: AXA.

Azizuddin, Khalid. 2020. "Germany to Assess Major European Firms Against EU Taxonomy." *Responsible Investor*, May 14. www.responsible-investor.com/articles/germany-to-assess-major-european-firms-against-eu-taxonomy.

Bolton, Patrick, Morgan Despres, Luiz Awazu Pereira da Silva, Frédéric Samama, and Romain Svartzman. 2020. *The Green Swan: Central Banking and Financial*

Stability in the Age of Climate Change. Basel: Bank of International Settlements and Bank of France.

Binham, Caroline, and Leslie Hook. 2019. "Markets Watchdog Seeks to Force Companies to Disclose Climate Risk." *Financial Times*, October 16. www.ft.com/content/837eea82-f02d-11e9-ad1e-4367d8281195.

China Banking and Insurance Regulatory Commission. 2012. *Guidance on Green Loans*. Beijing: China Banking and Insurance Regulatory Commission.

Climate Bonds Initiative. 2017. *Climate Bonds Standard, Version 2.1*. London: Climate Bonds Initiative.

Climate Bonds Initiative. n.d. "Green Bond Principles & Climate Bonds Standard." Accessed July 25, 2020. www.climatebonds.net/market/best-practice-guidelines.

European Commission. 2018. *Commission Action Plan on Financing Sustainable Growth*. Brussels: European Commission.

European Investment Bank and China's Green Finance Committee. 2017. *The Need for a Common Language in Green Finance, Towards a Standard-Neutral Taxonomy for the Environmental Use of Proceeds, China's Green Bond Endorsed Project Catalogue and the Common Principles for Climate Mitigation Finance Tracking Mapped and Compared*. Luxembourg: European Investment Bank and China's Green Finance Commit.

European Union Technical Expert Group on Sustainable Finance. 2019a. *Proposal for a European Union Green Bond Standard*. Brussels: European Union Technical Expert Group on Sustainable Finance.

European Union Technical Expert Group on Sustainable Finance. 2019b. *Report on EU Green Bond Standard*. Brussels: European Union Technical Expert Group on Sustainable Finance.

European Union Technical Expert Group on Sustainable Finance. 2019c. *Taxonomy Technical Report*. Brussels: European Union Technical Expert Group on Sustainable Finance.

European Union Technical Expert Group on Sustainable Finance. 2020a. *Taxonomy: Final Report of the Technical Expert Group on Sustainable Finance*. Brussels: European Union Technical Expert Group on Sustainable Finance.

European Union Technical Expert Group on Sustainable Finance. 2020b. *Taxonomy Report: Technical Annex*. Brussels: European Union Technical Expert Group on Sustainable Finance.

European Union Technical Expert Group on Sustainable Finance. 2020c. *Usability Guide EU Green Bond Standard*. Brussels: European Union Technical Expert Group on Sustainable Finance.

Federal Ministry for the Environment, Nature Conservation and Nuclear Safety. 2020. *Emerging Stronger from the Crisis Through Sustainable Economic Practices*. Bonn: Federal Ministry for the Environment, Nature Conservation and Nuclear Safety.

The Green Bond Principles and International Capital Market Association. 2018. *Voluntary Process Guidelines for Issuing Green Bonds*. Paris: The Green Bond Principles and International Capital Market Association.

Green New Deal, H.Res.109 (2019).

Ministry of the Environment. 2017. *Green Bond Guidelines, 2017*. Tokyo: Ministry of the Environment.

Ministry of the Environment. 2020. *Green Bond Guidelines 2020*. Tokyo: Ministry of the Environment.

Mishima, Daichi. 2019. "Japan's $1.5tn Pension Fund to Go All in on Green Bonds." *Nikkei Asian Review*, September 25. https://asia.nikkei.com/Spotlight/Environment/Japan-s-1.5tn-pension-fund-to-go-all-in-on-green-bonds.

Monetary Authority of Singapore. 2019. *Singapore: A Leading Centre for Green Finance in Asia and Globally*. Singapore: Monetary Authority of Singapore.

Network for Greening the Financial System. 2019. *A Call for Action, Climate Change as a Source of Financial Risk*. Paris: Network for Greening the Financial System.

Newsom, Gavin. Executive Order N-19–19 (2019).

People's Bank of China's Green Finance Committee of the China Society for Finance and Banking. 2015. *Green Bond Endorsed Project Catalogue*. Beijing: People's Bank of China's Green Finance Committee of the China Society for Finance and Banking.

Worland, Justin. 2020. "As the Rest of the World Plans a Green Recovery, America is Once Again Falling Behind." *Time*, May 15. https://time.com/5835402/green-stimulus-climate-change-coronavirus/.

6 Scenario analysis

Scenario analysis for assets can help investors identify future carbon risks and green finance opportunities.

Since the global financial crisis of 2008, banks have increasingly used scenario analysis. Certain financial supervisors, such as the Bank of England, have been requiring banks to use this analytical tool to evaluate how specific scenarios would impact the assets in their investment portfolios. Financial supervisors themselves have also used this analytical tool to evaluate banks operating within their jurisdiction. The general purpose of this has been to determine whether a major financial crisis, such as the economic conditions associated with a severe recession, would adversely impact a bank's ability to satisfy its capital obligations with respect to its lending or face insolvency. Insurers have also been using scenario analysis to evaluate their ability to pay insurance claims made on premiums (Dent, Westwood, and Segoviano 2016).

Scenario analysis is one of the global financial community's most sophisticated analytical tools for evaluating financial risks. This tool is now being used to identify future carbon risks and green finance opportunities. Scenarios used in analyzing assets for transition risks are directly or indirectly based on energy technology deployments and other activities deemed necessary in order to achieve greenhouse budgets connected with temperature targets. The creation of these scenarios has been made possible through advancements with monitoring greenhouse gases and determining their global warming potential, both with respect to the development of an overall budget and with respect to identifying the emissions associated with different sectors of the economy. Scenarios used in analyzing assets for physical risks tend to be based on emissions trajectories, which efforts to monitor greenhouse gases have made possible. The ability to analyze companies' investment portfolios and provide guidance on future action is the result of advancement with tracking investments, in particular with linking securities to fossil fuel and sustainable classifications (IEA 2018; Greenpeace n.d.; IREA n.d.; PRI n.d.b).

6.1 Transition risks

As a tool already used in the financial sector for other purposes, the TCFD recommended using scenario analysis to test climate scenarios on assets under management as a principal means for evaluating carbon risks, in particular for transition risks associated with a shift to a low-carbon economy. In addition to having significant credibility because of its already existing application, the use of scenario analysis has garnered much positive attention. This is because it allows for analysis that has not been possible with other analytical tools. In particular, scenario analysis allows for forward-looking analysis, whereas other analytical tools assess currently available information. Conducting forward-looking analysis is crucial because although the climate is currently impacting markets and properties, some of the most significant changes in the global economy and with property are not expected to take place until the future. Researchers further believe that it is important to evaluate these changes in order to plan for them.

Following the TCFD recommendations, companies are increasingly running climate-related scenario analysis tests on the assets in investment portfolios. The most common of these tests help companies determine the exposure of their assets to a transition away from fossil fuel investments toward sustainable investments. According to the TCFD:

> A key type of transition risk scenario is a so-called 2°C scenario, which lays out a pathway and an emissions trajectory consistent with holding the increase in the global average temperature to 2°C above pre-industrial levels. In December 2015, nearly 200 governments agreed to strengthen the global response to the threat of climate change by "holding the increase in the global average temperature to well below 2°C above pre-industrial levels and to pursue efforts to limit the temperature increase to 1.5°C above pre-industrial levels," referred to as the Paris Agreement. As a result, a 2°C scenario provides a common reference point that is generally aligned with the objectives of the Paris Agreement and will support investors' evaluation of the potential magnitude and timing of transition-related implications for individual organizations; across different organizations within a sector; and across different sectors.
>
> (TCFD 2017a, 27)

Also, according to the TCFD, the International Energy Agency (IEA) energy technology deployments scenarios are the scenarios that are typically used to evaluate transition risks (TCFD 2017a, 12). The IEA has been using a World Energy Model to make energy projections for decades. Updated every year, the model is composed of three primary modules: final energy consumption; energy transformation, including power generation and heat; and energy supply. The IEA has used the World Energy Model to generate sector-by-sector and region-by-region energy technology deployment scenarios (IEA 2018).

Initial energy technology deployment scenarios that were made by the IEA included its 450 Scenario, New Policies Scenario, and Current Policies Scenario. The IEA's 450 Scenario identifies how the IEA believes energy technologies should be deployed in order to limit global warming to 2 degrees Celsius. The IEA's New Policies Scenario identifies how the IEA believes energy technologies will be deployed if all new policies set out in Nationally Determined Contributions are implemented to meet countries commitments under the Paris Agreement. The IEA's Current Policies Scenario identifies how the IEA believes technologies will be deployed with business as usual without new climate policies (IEA 2018; PRI n.d.b; Bolton et al. 2020, 36–41). The initial IEA energy technology deployment scenarios, which continue to be updated over time, are shown in Table 6.1.

As international climate objectives and social preferences have continued to advance, the IEA has also created supplemental energy technology deployment scenarios. This includes the IEA's Sustainable Development Scenario, which shows the energy technologies that should be deployed by 2050 in order to reach a fifty percent probability of limiting global warming to 2 degrees Celsius, but the scenario also incorporates social responsibility factors when deciding on the appropriate deployment of energy technologies. This also includes the IEA's Beyond 2 Degrees Scenario, which reflects the interest in below 2 degrees Celsius modeling following the IPCC and United States National Climate Assessment reports that stressed the need to achieve the below 2 degrees goals of the Paris Agreement, rather than simply 2 degrees Celsius targets. It is more difficult for the investment portfolios of companies to be aligned with this scenario, as it requires more aggressive deployment of sustainable energy technologies (IEA 2018; CDI and 2DII 2018, 28–29). With a current emphasis on achieving a 1.5 degrees Celsius

Table 6.1 Initial Scenarios for Transition Risks

IEA's 450 Scenario	The International Energy Agency's 450 Scenario identifies how the International Energy Agency believes energy technologies should be deployed in order to limit global warming to 2 degrees Celsius.
IEA's New Policies Scenario	The International Energy Agency's New Policies Scenario identifies how the International Energy Agency believes energy technologies will be deployed if all new policies set out in Nationally Determined Contributions are implemented to meet countries' commitments under the Paris Agreement.
IEA's Current Policies Scenario	The International Energy Agency's Current Policies Scenario identifies how the International Energy Agency believes technologies will be deployed with business as usual without new climate policies.

target, the use of below 2 degrees Celsius scenarios is likely to increase moving forward. See Table 6.2.

Additional scenarios exist, too, such as Greenpeace's Advanced Energy [R]evolution and the International Renewable Energy Agency's REmap scenarios. Use of these scenarios could be desired for various reasons but predominantly because a user may wish to examine the impact of a scenario that makes different assumptions about how energy technology deployment will develop moving forward. For instance, while the IEA scenarios are the result of significant consultation efforts, they rely on assumptions for energy projections decades into the future, which could be contested. In particular, the IEA has historically assumed a significant amount of nuclear power and carbon capture and storage deployment in their scenarios. However, other scenarios may not have been constructed with the belief that nuclear power will be used in the future because of regulatory practices or consumer interests, especially outside of regions that are not already heavily invested in the technology. Likewise, other scenarios may not have been constructed with the belief that carbon capture and storage will offer promising emissions reductions moving forward. Greenpeace's Advanced Energy [R]evolution scenario, for instance, does not take into account either nuclear power or carbon capture and storage and shows what needs to be achieved in order to reach a fully decarbonized energy system by 2050 (Greenpeace n.d.; PRI n.d.b). International Renewable Energy Agency's REmap envisions an energy sector that rapidly transitions to clean technology, with a doubling of the share of energy coming from renewable sources by 2030 (IREA n.d.; PRI n.d.b). It will likely be more difficult for companies to be aligned with Paris Agreement objectives when comparing their investment portfolios to Greenpeace's Advanced Energy [R]evolution pathways in comparison to IEA pathways. Alignment with International Renewable Energy Agency's REmap scenario is also likely to be more difficult unless a company is heavily invested in renewable energy. See Table 6.3.

Table 6.2 Supplemental Scenarios for Transition Risks

IEA's Sustainable Development Scenario	The International Energy Agency's Sustainable Development Scenario identifies how the International Energy Agency believes energy technologies should be deployed in order to limit global warming to 2 degrees Celsius while taking into account social responsibility factors.
IEA's Beyond 2 Degrees Scenario	The International Energy Agency's Beyond 2 Degrees Scenario identifies how the International Energy Agency believes that energy technologies should be deployed in order to limit global warming to 1.75 degrees Celsius.

Table 6.3 Additional Scenarios for Transition Risks

Greenpeace's Advanced Energy [R]evolution	Greenpeace's Advanced Energy [R]evolution envisions a fully decarbonized energy system.
International Renewable Energy Agency's REmap	International Renewable Energy Agency's REmap envisions a rapid doubling of the renewable energy share of the world's energy mix.

6.1.1 Energy transition

Today, a number of financial supervisors are using scenario analysis to evaluate transition risks, including the Bank of England, the Bank of France, and the Dutch Bank (Bank of England 2019; Bank of France 2020; Vermeulen et al. 2018, 2019; NGFS 2020). Initial widescale application of transition risk scenario analysis on assets in investment portfolios took place in 2017 and 2018. In 2017 and 2018, the Swiss and Californian governments conducted extensive analyses of the investment portfolios of companies operating in their jurisdictions using the Paris Agreement Capital Transition Assessment (PACTA) scenario analysis tool developed by 2 Degrees Investing Initiative with funding from the European Union. The Swiss Ministry of Environment and 2 Degrees Investing Initiative effort offered the tool to all Swiss pension funds and insurers to test the climate alignment of their equity and corporate bond portfolios on a voluntary basis. It was possible to analyze two-thirds of the assets under management (2DII 2017; Thomä and Chenet 2019) and in doing so identify potential transition risks in portfolios. The California Department of Insurance worked with 2 Degrees Investing Initiative to apply the tool to investment portfolios of several hundred of the largest insurers operating in California and identified potential transition risks there. The analysis specifically assessed whether investments were aligned with emissions pathways consistent with holding the increase in global average temperature to 2 degrees Celsius above preindustrial levels (CDI and 2DII 2018, 12–17).

The PACTA tool allows a company or other organization to measure, monitor, and address the alignment of an investment portfolio with climate objectives (2DII n.d.c). This type of tool is an energy transition scenario analysis tool where securities in companies' investment portfolios are matched against energy technology deployment needed to meet desired temperature targets for specified dates. An energy transition scenario analysis tool uses climate scenarios such as those developed by the IEA. The IEA, for instance, develops energy transition scenarios by making global projections for energy technology deployment through assembling information from national and subnational energy regulatory plans. The scenarios can be further calibrated for global market or regional market designations, such as the countries making up the Organization for Economic Co-operation and Development (OECD) (CDI n.d.).

For the initial analysis that the California Department of Insurance and 2 Degrees Investing Initiative performed with the PACTA tool, the energy technology deployment plans that were used were made by the IEA and included its 450 Scenario, New Policies Scenario, and Current Policies Scenario. The IEA's 450 Scenario, the most stringent scenario, was used here to show how energy technologies should be deployed by 2050 in order to reach a fifty percent probability of limiting global warming to 2 degrees Celsius. The IEA's New Policies Scenario, a more moderate scenario identifying how the IEA believes energy technologies will be deployed if all new policies set out in Nationally Determined Contributions are implemented to meet countries commitments under the Paris Agreement, was used here to show how energy technologies should be deployed by 2050 in order to reach a fifty percent probability of limiting global warming in accordance with Nationally Determine Contributions. The IEA's Current Policies Scenario, the least stringent scenario identifying how the International Energy Agency believes technologies will be deployed with business as usual without new climate policies, was used here to show how energy technologies should be deployed by 2050 in order to reach a fifty percent probability of limiting global warming in accordance with business as usual (CDI and 2DII 2018, 24–25). The scenarios were calibrated for OECD countries, which is where the insurers and most of the companies that they have been invested in are based (CDI n.d.).

Pathways for the IEA scenarios were mapped out on a chart over a five-year span. There is green shading below the IEA's 450 Scenario pathway that identifies an area where there is likely alignment with 2 degrees Celsius energy technology deployment when analyzing fossil fuel enterprises. There is light yellow shading between the IEA's 450 Scenario and the IEA's New Policies Scenario pathways that identifies an area where there is likely some misalignment with 2 degrees Celsius energy technology deployment and instead likely alignment with 2 degrees Celsius energy technology deployment to Nationally Determined Contributions energy technology deployment. There is dark yellow shading between the IEA's New Policies Scenario and the IEA's Current Policies Scenario pathways that identifies an area where there is also likely some misalignment with the 2 degrees Celsius energy technology deployment and instead likely alignment with Nationally Determined Contributions energy technology deployment to business-as-usual energy technology deployment. Further, there is red shading above the IEA's Current Policies Scenario that identifies an area where there is likely significant misalignment with 2 degrees Celsius energy technology deployment and instead likely alignment with beyond business-as-usual energy technology deployment (CDI and 2DII 2018, 14–17).

The PACTA tool takes into account a company's inputs of both equity and debt investments and uses that information to generate a comprehensive report consisting of several dozen pages. The report evaluates alignment for several industry sectors, such as coal, oil, gas, utilities generating electric power, renewable energy, and transportation. For each sector, there is a line for equity and debt investments that shows the trajectory of those investments over the five-year time horizon against the green, light yellow, dark yellow, and red background. Again, when a line is in a green area, this indicates that a company's investments are likely aligned with energy technology deployment necessary to limit temperature increases to 2 degrees Celsius. When the line is in a light or dark yellow area, this indicates that a company's investments are likely somewhat in misalignment with energy technology deployment necessary to limit temperature increases to 2 degrees Celsius, to varying amounts. When a line is in a red area, this indicates that a company's investments are likely significantly misaligned with energy technology deployment necessary to limit temperature increases to 2 degrees Celsius (CDI and 2DII 2018, 14–17).

Figure 6.1 shows a sample of investment portfolio trajectories transposed on pathway data. Here the solid line represents aggregate capacity for equity within the investment portfolios of insurers operating in California. Figure 6.1 specifically examines the trajectory of investments in utilities operating on the burning of thermal coal.

Because the solid line representing the trajectory of the equity within the investment portfolios falls within the red area, which in the grayscale reproduction here appears as a dark gray area at the top of the figure, this shows

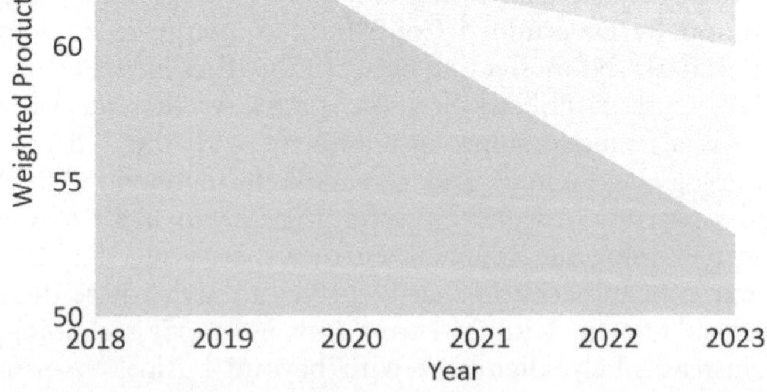

Figure 6.1 Utilities Generating Electricity From Coal

Source: 2 Degrees Investing Initiative

that these investments are likely significantly misaligned with the energy technology deployment necessary to limit temperature increases to 2 degrees Celsius. In order for investments in these utilities to be aligned with energy technology deployment necessary to limit temperature increases to 2 degrees Celsius and for the line representing the trajectory of the equity within the investment portfolios to therefore fall in the green area, which in the gray-scale reproduction here appears as a lighter gray area at the bottom of the figure, the utilities that are being invested in would need to massively retire coal power plants (CDI and 2DII 2018, 15; Mills et al. 2018, 27–29).

While findings varied across companies, published aggregate data indicated that overall, there was concern that too much money was being invested in certain fossil fuel industries such as thermal coal and not enough in renewable energy or other clean energy technologies. On a positive note, insurer equity and debt investments were better aligned with a 2 degrees Celsius scenario than the market as a whole. Moreover, there was not significant misalignment with oil and gas production (CDI and 2DII 2018, 15; Piotrowski 2018).

While aggregate information released by the Swiss and Californian governments did not reveal security information, specific asset risk information was provided confidentially in individual reports that were provided directly to companies. These individual reports showed sectors where companies were most exposed to transition risks with charts specifically tailored to the investment trajectories of individual companies (CDI and 2DII 2018, 12–19; Piotrowski 2018). Moreover, at the end of reports, investment information was provided that identified specific securities that are most adversely impacting the investment trajectories of individual companies, that is, pushing an investment trajectory line into a yellow or red area (CDI and 2DII 2018, 20–21). This information could be used by companies to align their portfolios with the objectives of the Paris Agreement and for French Energy Transition Law or TCFD reporting.

The PACTA tool can also use other scenarios that have been developed as international climate objectives and social preferences have changed over time or been adjusted to reflect updated scientific research. For example, the California Department of Insurance and 2 Degrees Investing Initiative decided to apply the IEA's Sustainable Development Scenario when applying the PACTA tool to the investment portfolios of insurers in a subsequent year of analysis rather than the IEA's 450 Scenario (CDI and 2DII 2018, 28–29; CDI n.d.). The California Department of Insurance and 2 Degrees Investing Initiative also added the IEA's Beyond 2 Degrees Scenario when applying the PACTA tool to the investment portfolios of insurers in the subsequent year of analysis. In this subsequent analysis, the California Department of Insurance and 2 Degrees Investing Initiative continued to use the IEA's New Policies Scenario and the IEA's Current Policies Scenarios. However, when this analysis was conducted, temperature projections associated with these scenarios had decreased (CDI and 2DII 2018, 28–29; CDI n.d.).

2 Degrees Investing Initiative launched a free-to-use online version of the PACTA tool. The tool allows analysis of exposure in transition risks in equity and debt investments within portfolios over multiple scenarios (PRI n.d.a). This allows users to manipulate criteria such as temperature targets or energy technology deployments for achieving those targets. It also allows for applying additional scenarios, such as Greenpeace's Advanced Energy [R]evolution and the International Renewable Energy Agency's REmap scenarios.

Application of the PACTA tool has grown. 2 Degrees Investing Initiative now has had over a thousand companies upload their investment portfolio into the free-to-use online version of its tool (2DII n.d.c). The PACTA tool also became one of the tools for the Climate Action 100+, which is used by investors to influence more than a hundred of the largest emitters of greenhouse gases (Climate Action 100+ 2019, 76). ING launched Terra, which uses PACTA technology to evaluate the ING's investment portfolio's exposure to transition risks (ING 2020). The European Insurance and Occupational Pensions Authority has been using PACTA to examine the portfolios of European insurers (2DII n.d.b). The Swiss Ministry of Environment is also using PACTA for an updated scenario analysis effort that includes participants from additional sectors, such as banking, and covers more types of assets (2DII n.d.a).

A transition risk tool that uses energy transition scenarios allows for the analysis of a significant range of scenarios, providing a user with a great degree of flexibility in analyzing assets under management. An energy transition scenario can also inform policy decisions on the best approaches to managing energy consumption moving forward with respect to decisions on deployment of energy technologies as a share of the energy mix in short-, medium-, and long-term timeframes. Following the signing of the Paris Agreement, 2 degrees Celsius was a target that was often used, as it was decided that a more than 2 degrees Celsius rise would lead to catastrophic carbon risks. The California Department of Insurance and 2 Degrees Investing Initiative used several scenarios when initially applying the PACTA tool, with an emphasis on the IEA's 450 Scenario looking at limiting global warming to 2 degrees Celsius. This was the most widely used transition risk scenario at the time (CDI and 2DII 2018, 24), as the world was focused on achieving the Paris Agreement objective of limiting global temperature rise to no more than 2 degrees Celsius above preindustrial levels. However, following the IPCC and United States National Climate Assessment's recent bleak assessments of climate change impacts, now 1.5 degrees Celsius is becoming a widely accepted target. There has also been a push to include the management of additional socially responsible investment standards with the management of environmentally responsible standards. This has been reflected in an expansion in the capabilities of transition risk scenario analysis tools. Therefore, these tools, like those used for the subsequent California Department of Insurance and 2 Degrees Investing Initiative scenario analysis, are able to incorporate the IEA's Beyond 2 Degrees and Sustainable Development scenarios. There is also increased debate over the appropriate energy technology deployment projections, with some projections taking into account nuclear energy and carbon capture and storage and others not doing so.

6.1.2 Warming potential

Another type of transition risk tool that has evolved is the warming potential scenario analysis tool. This tool compares the global warming potential of securities in a company's investment portfolio against total greenhouse gas emissions allowable in order to reach specified temperature targets. Carbon Delta offers a warming potential tool called Climate Value-at-Risk that operates in the following manner. To begin with, Scope 1 and Scope 2 greenhouse gas budgets are established for each sector and for companies within those sectors. These budgets are based on projections necessary to achieve certain temperature targets. Carbon Delta then determines necessary emissions reductions for companies at the facility level in order to meet these budgets, using a proprietary database of over twenty thousand publicly-listed companies covering equity and debt for over tens of thousands of securities (MSCI and Carbon Delta 2019).

Carbon Delta's tool is available commercially on a subscription or project basis (MSCI and Carbon Delta 2019). In a TCFD and French Energy Transition Law report, AXA stated that it is examining using Carbon Delta's tool to measure the position of its portfolio versus a 2 degrees Celsius target (AXA 2018, 8–11). AXA is now using Carbon Delta's tool to test climate scenarios, including a 1.5 degrees Celsius scenario, on the value of its investments (AXA 2020, 16). MSCI, one of the world's largest data providers, recently acquired Carbon Delta (MSCI n.d.).

A warming potential scenario analysis tool provides some unique capabilities when assessing climate change specific to the operations of individual companies. In particular, such a tool could eventually incorporate Scope 3 emissions, providing the opportunity to examine supply chain contributions in ways that other tools are not built to easily do. Such tools also provide unique challenges, though. Challenges include, for instance, how reliable and comparable greenhouse gas measurements are and whether the information is independently verified. Right now, there are possible gaps in coverage, as there are not a lot of independently verified carbon footprints for all emissions scopes, but coverage is likely to increase over time.

6.1.3 Carbon price

Another type of transition risk scenario analysis tool that has evolved is the carbon price scenario analysis tool. Trucost, which is now part of S&P Global (Mattison 2018), has developed a carbon price scenario analysis tool called Carbon Earnings At Risk. According to research cited by Trucost, significantly increased carbon pricing will be necessary in order to address climate change, a several-fold increase to over one hundred dollars per ton of CO_2 (Trucost and S&P Global 2019a, 2019b, 2019c).

Carbon Earnings At Risk compares the potential implementation of carbon price mechanisms to companies' investment portfolios based on the potential application of carbon price mechanisms in specific regions where companies in the portfolio are located and on expected carbon emissions for these companies over time. To achieve this aim, Trucost has developed several scenarios that can be applied to companies' investment portfolios to evaluate the impact of an increase in carbon pricing. Using a database of present and potential carbon pricing mechanisms, such as carbon taxes and emissions trading systems, Trucost has developed scenarios that include projected carbon pricing mechanisms needed to achieve temperature projections connected with the IEA's 450 Scenario and the New Policies Scenario. Trucost has also developed a database of greenhouse gas emissions associated with particular securities using publicly available greenhouse gas data provided to CDP (Trucost and S&P Global 2019a, 2019b, 2019c).

A growing number of companies are reporting to CDP that they are pricing carbon internally to evaluate their own operations (Trucost and S&P Global 2019a, 2019b, 2019c). Along these lines, this tool allows investors to evaluate the impact of carbon price increases on their portfolio of investments. As application of the tool is centered around specific regulatory actions, such as the pricing of carbon through further implementation of carbon taxes or emissions trading systems, a carbon price tool can inform a company of impacts that are likely to occur if a large policy decision is made on pricing carbon and also inform policymakers on the impact of such a decision.

6.1.4 Capex

In addition to transition risk scenario analysis that examines transition risks on investments, there are models that look at transition risks with fossil fuel reserves, in particular for extensions to existing projects or for new projects where there has not been a final investment decision. A prominent transition risk tool that provides this type of analysis is the 2 Degrees of Separation tool developed by Carbon Tracker and PRI. This tool looks at the upstream risks to the oil and gas sectors and oil and gas companies and projects. It can be used to determine whether companies are highly exposed to transition risks based on capital expenditures or capex necessary to exploit their oil and gas reserves. The general premise of the tool is that not all oil and gas companies are equally positioned for a transition to a low-carbon economy. A company is not aligned with the Paris Agreement objectives if it plans to exploit reserves in excess of emissions limits necessary to achieve these objectives. Exploitation of these reserves is also not necessarily financially viable unless the market price of oil and gas is exceedingly high. Companies with reserves that have a high extraction cost should be advised that a restriction on the use of oil and gas and related or unrelated decreases in consumer demand will most adversely impact those companies with the highest extraction cost for oil and gas reserves, whereas companies with lower extraction costs for oil and gas reserves will still be viable, at least in the near term (Leaton and Grant 2017).

When the 2 Degrees of Separation tool launched in 2017, the tool used the IEA's 450 Scenario to evaluate oil and gas sectors and oil and gas companies and projects. In particular, the tool developed oil and gas carbon budgets for 2017 to 2035 using the IEA's 450 Scenario as a proxy for Paris Agreement objective of limiting average global temperature to no more than a 2 degrees Celsius increase above preindustrial levels. This allows the tool to identify the use of fossil fuels that will produce no more than a set limit of carbon dioxide emissions, split fifty-nine percent for oil and forty-one percent for gas. By plotting a vertical carbon budget line on a supply cost curve chart, limits for both oil and gas can be compared to a business-as-usual supply cost curve sourced from industry data. The supply cost curve to the left of a vertical carbon budget line is perceived as within the 2 degrees Celsius budget, the supply cost curve to the right of a vertical carbon budget line is perceived as outside the 2 degrees Celsius budget, and connected projects are viewed as involving capex that should not be deployed if the world were to align with the Paris Agreement objectives (Leaton and Grant 2017, 15–17).

The tool provides sectorwide and company- and project-specific information. Overall, for the oil and gas sectors, the tool determined that 2.3 trillion dollars, or approximately one-third of potential capex to 2025, should not be deployed if the world were to align with the Paris Agreement objectives (Leaton and Grant 2017, 5, 19–22, 25). Figure 6.2 shows the application of the 2 Degrees of Separation tool for the oil sector. The supply cost curve to the left of the vertical oil emissions budget line is perceived as within the 2 degrees Celsius budget, the supply cost curve to the right of the vertical oil emissions budget line is perceived as outside the 2 degrees Celsius budget, and connected projects are viewed as involving capex that should not be deployed if the world were to align with the Paris Agreement objectives. Projects to the right of the vertical emissions budget line, on the steeper side of the supply cost curve, include oil sands or certain deepwater projects. Typically, the highest-costing projects are also the most carbon-intense projects (Leaton and Grant 2017, 19).

The tool also was used to generate a list showing the percentage of total available capex opportunities for specific oil and gas companies' potential projects. The companies with the least capex outside of the 2 degrees Celsius budget were viewed as facing the least risk if the world were to align with the Paris Agreement objectives. On the other hand, those with the most capex outside the 2 degrees Celsius budget were viewed as facing the most risk (Leaton and Grant 2017, 26–29).

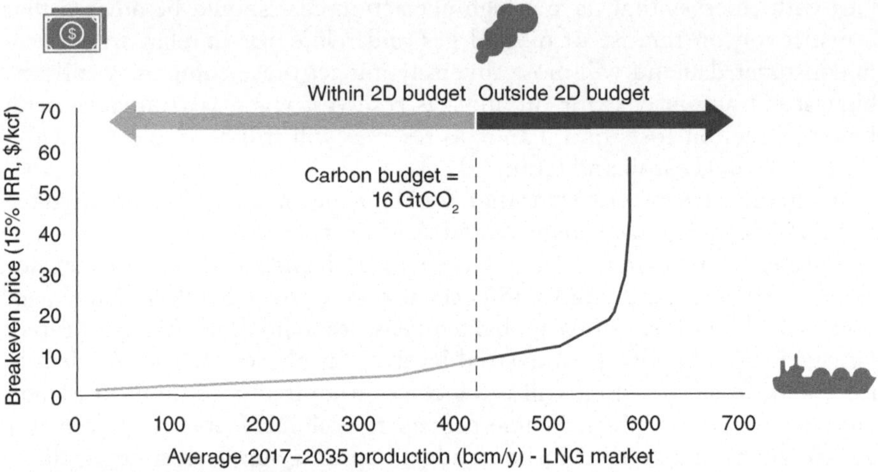

Figure 6.2 Oil Production Cost Curve

Source: Carbon Tracker. This image was viewable on July 25, 2020, on the following Carbon Tracker website: www.carbontracker.org/reports/2-degrees-of-separation-transition-risk-for-oil-and-gas-in-a-low-carbon-world-2/

In 2019, additional work with the 2 Degrees of Separation tool applied the IEA's New Policies Scenario, an updated IEA's Sustainable Development Scenario and an IEA's Beyond 2 Degrees Scenario on the oil and gas sectors for years 2019 to 2030. Under the IEA's New Policies Scenario, the tool does not identify a risk of stranding capital expenditures. However, under an updated IEA's Sustainable Development Scenario, the tool identifies a risk that sixty percent of capital expenditures will be stranded. Further, under an IEA's Beyond 2 Degrees Scenario, the tool identifies a risk that eighty-three percent of capital expenditures will be stranded. The tool further identifies oil and gas companies that are invested in nineteen major projects that are not aligned with the IEA's Sustainable Development Scenario. According to the tool's application, there is fifty billion dollars of investment in these nineteen major projects that are inconsistent with a Sustainable Development Scenario pathway (Carbon Tracker 2019). Carbon Tracker and PRI have made 2 Degrees of Separation reports public and have provided additional analytical resources to PRI signatories (PRI n.d.b).

The capex scenario analysis tool provides useful information to a number of parties. For fossil fuel companies, the tool can help inform companies on the most cost-effective reserves to exploit and whether additional exploration is desirable. The tool can help investors understand which companies are most exposed to a transition to a low-carbon economy. The tool can also help policymakers understand the supply cost pressures on fossil fuel companies when balancing other considerations such as geopolitical and environmental concerns.

6.1.5 Comparing tools

There are a number of different types of transition risk scenario analysis tools. These tools serve different functions. Tables 6.4 and 6.5 discuss the transition risk scenario analysis tools for investments and fossil fuel reserves, respectively.

Table 6.4 Transition Risk Scenario Analysis Tools for Investments

Energy Transition	An energy transition scenario analysis tool compares investment portfolios to energy transition scenarios that are estimated to be needed in order to reach a temperature target.
Warming Potential	A warming potential scenario analysis tool compares investment portfolios to global warming budgets that are needed in order to reach a temperature target.
Carbon Price	A carbon price scenario analysis tool examines the impact of a carbon price (e.g., resulting from a cap-and-trade program or carbon tax) on investment portfolios.

Table 6.5 Transition Risk Scenario Analysis Tools for Fossil Fuel Reserves

Capex	A capex scenario analysis tool evaluates whether potential capex should be deployed when considering carbon budget restrictions.

6.2 Physical risks

Physical risk scenario analysis is increasing in prominence. Physical risk scenario analysis involves using climate scenarios to evaluate the physical risks presented to assets under management. The IPCC's climate scenarios are the scenarios typically used to evaluate physical risks (TCFD 2017b, 12–13). Incorporating the input of governments and research institutions involved in dozens of global research collaborations (IPCC 2020; PRI n.d.b; Four Twenty Seven 2018, 2–3), the scenarios identify emissions trajectories or the amount of emissions in the atmosphere over time when specific future courses of action are taken (IPCC 2020; PRI n.d.b; TCFD 2017b, 24–25). In Table 6.6 is information on IPCC's climate scenarios that are used for physical risk ratings and scenario analysis that incorporates physical risk ratings.

Table 6.6 Initial Scenarios for Physical Risks

IPCC's RCP 2.6	The Intergovernmental Panel on Climate Change's Representative Concentration Pathway 2.6 Scenario is a scenario where there is strict mitigation action and global temperatures are aligned with Paris Agreement objectives.
IPCC's RCP 4.5	The Intergovernmental Panel on Climate Change's Representative Concentration Pathway 4.5 Scenario is an intermediate mitigation action scenario where there are sizeable global temperature increases.
IPCC's RCP 8.5	The Intergovernmental Panel on Climate Change's Representative Concentration Pathway 8.5 Scenario is a business-as-usual scenario where there are significant global temperature increases.

6.2.1 Generating ratings

The general methodology for physical risk scenario analysis builds on practices already established for rating physical risks to properties. Insurers and banks have been using property-specific risk scores, generated from risk assessment tools, to make lending and underwriting decisions about specific properties, while total overall underwriting rates can be restricted by actuarial models that use historic loss data. In the United States, the primary risk assessment tool for wildfires has been generating property-specific risk scores using historical data with only three factors: fuel load, access, and slope for all properties within a zip code. Properties in specific zip codes are assigned a score of 0 to 30 based on a calculation for their zip code, with a score of 0 representing no risk and a score of 30 representing extreme risk (ISO n.d.). Insurers will typically charge lower premium rate to properties with a low risk score and charge a higher rate to properties with a high risk score or decide not to underwrite a property with high risk score altogether.

A risk assessment tool that is capable of generating property-specific risk scores using a range of complex environmental factors can also assign risk scores using forward-looking weather and non-weather data sourced from downscaled climate model output. Climate models can provide information on dozens of variables, including temperature and precipitation, but climate models produce gridded data with limited resolution. However, climate model output can be further downscaled to provide greater resolution, allowing for modeling on topographical features (Four Twenty Seven 2018, 2–3). Once downscaled, climate change model output can be supplemented with additional data to yield more location specific risk assessment information. This can be loaded into a risk assessment tool to provide property-specific risk scores.

The process for developing forward-looking risk rating involves selecting a climate scenario to input into a climate model. A climate modeler can input climate scenarios, such as the IPCC's scenarios, into climate models to generate outputs. For instance, as part of California's Fourth Climate Assessment, Scripps Institution of Oceanography researchers input the IPCC's RCP 4.5 Scenario and the IPCC's RCP 8.5 Scenario into global climate models. The output from doing so was limited in resolution. However, the Scripps Institution of Oceanography researchers then used localized constructed analogs to downscale the limited-resolution outputs. Doing so increased resolution, allowing for modeling on topographical features such as California's unique land surface and hydrological characteristics (Pierce, Cayan, and Dehaan 2016). Once downscaled, climate change model projections can be supplemented with additional data, including, for instance, heat stress, water stress, wildfires, sea level rise, and hurricanes, to yield more location-specific risk assessment information (Four Twenty Seven 2018, 4–7). Downscaling is depicted in Figure 6.3.

Four Twenty Seven, an affiliate of Moody's and provider of data related to physical climate and environmental risks (Flavelle 2019), uses its risk assessment tool to provide physical climate risk scores based on downscaled global climate models and other environmental datasets applied to corporate structures and other property assets (Four Twenty Seven 2018, 7). This process has enabled Four Twenty Seven to develop a global database of security information with facility-level physical risk scores (Four Twenty

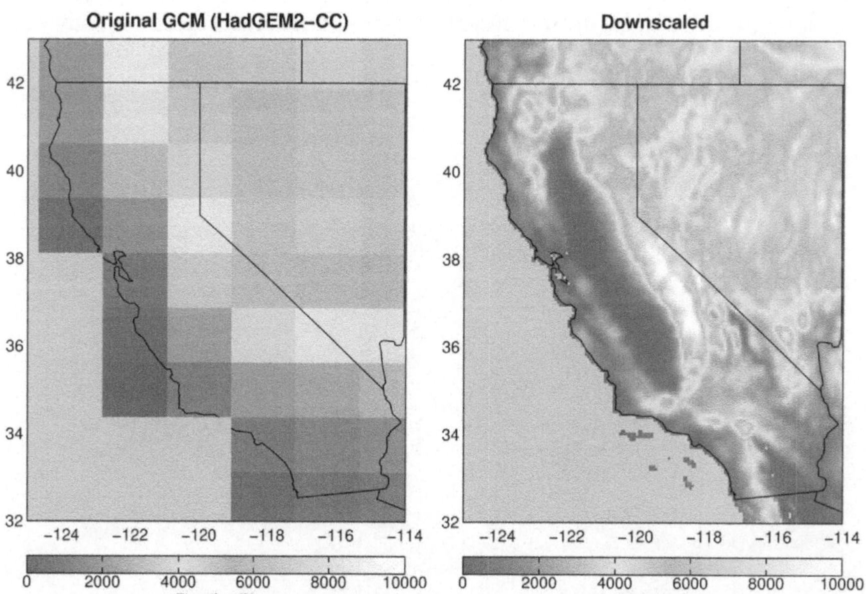

Figure 6.3 Climate Model Downscaling

Source: Scripps Institution of Oceanography

Seven 2018, 7). Figure 6.4 shows the exposure to water stress of all facilities in Four Twenty Seven's global company database. The map provides overall risk scores for corporate facilities and other property holdings. Although not as easily discernable in the grayscale reproduction, distinguishable shading represents locations where there is the highest level of risk, significant risk, still concerning risk, and minimal risk. Where there is no shading, the database did not include facilities in that area at the time this map was generated, although the database's coverage has since expanded.

Four Twenty Seven's on-demand physical climate risk application and company climate risk scores allow asset owners to evaluate identifiable climate risks and determine whether any risk mitigation efforts are warranted. For instance, asset owners can assess flood risks for a particular asset and then explore whether there are specific protection efforts that should be taken or whether relocation efforts are more prudent. With heat stress, asset owners can determine how increased energy demands could impact operational costs and plan for the impact accordingly (Four Twenty Seven 2019, 13).

Four Twenty Seven is also planning to add additional scenarios to its physical climate risk scores. While Four Twenty Seven's database already allows asset managers to check their portfolio against Four Twenty Seven's database of security information to determine how exposed they are to physical risks, additional RCP scenarios would help inform reporting requirements and risk management. The knowledge can inform future asset management decisions in order to be more aligned with the Paris Agreement objectives.

Figure 6.4 Water Stress Risks

Source: Four Twenty Seven (427mt.com)

6.2.2 Applying ratings

A physical scenario analysis tool can incorporate physical risk ratings in the following manner. A physical scenario analysis tool links property-specific risks scores to corporate assets based on the location of the company's business operations. A modeler uploads investment portfolio information into this physical scenario analysis tool, which provides information on the exposure of assets in that portfolio to the risks reflected in the risk scores.

In the beginning of 2019, when the California Department of Insurance and 2 Degrees Investing Initiative released subsequent scenario analysis results, the results included an assessment of physical risks on insurers' investments, along with transition risks (CDI n.d.). In particular, the analysis looked at water stress, floods, and wildfires. With respect to water stress, the analysis used World Resource Institute's Aqueduct risk rating tools and data to assess the investments under management and determined the percentage of thermal power investments at risk of water stress. Thermal power plants, including coal, oil, gas, and nuclear power plants require water for operations such as cooling and are at risk from climate-related water stress. The data also indicates that climate change presents risks to certain clean technologies exposed to water stress, such as thermal solar power plants in desert regions (CDI and 2DII 2019, 23; WRI n.d.). With respect to flood risk, the analysis applied the European Union–developed global flood risk rating map to identify the exposure of power, oil and gas, and coal mining investments to a 1-in-100 scenario (CDI and 2DII 2019, 24; EU 2020). With respect to wildfire risks, the analysis applied wildfire risk analytics built on the United Nations Environment global wildfire risk rating map to identify the exposure of power, oil and gas, and coal mining investments. The data also indicates that climate change presents risks to utilities with transmission lines in certain regions where there is significant wildfire risk (CDI and 2DII 2019, 23). All of the physical risk analysis used the underlying investment portfolio data used for the subsequent transition risk analysis. Table 6.7 discusses how a physical risk scenario analysis tool for assets works.

Physical risk scenario analysis is important because it helps asset managers understand another important concern when making decisions about investments. Some investments that are not exposed to transition risks could be exposed to physical risks. Moreover, other investments may be exposed to both transition and physical risks, requiring additional attention.

Table 6.7 Physical Risk Scenario Analysis Tool for Assets

Location of Assets	This scenario analysis uses a physical risk rating or other location-specific assessments that are attributed to securities through their headquarters and other real property holdings.

6.3 Comparing risks

Transition scenario analysis and physical risk scenario analysis have important similarities. To start, with both transition and physical risk scenario analysis, users of scenario analysis have to select scenarios. Scenarios incorporate a number of assumptions that are unique to each scenario. However, the main decision that a user has to make in selecting a scenario is how ambitious future action will be with respect to achieving climate objectives with scenarios reflecting either major, moderate, or no achievements. In addition, there are also subtle decisions that a user has to make in selecting a scenario, such as whether a modeler agrees with data that was used in developing the scenario and how that data was incorporated.

Next, with both transition and physical risk scenario analysis, users have to create or decide on how to upload investment portfolio data. These decisions impact how to compose and otherwise aggregate the data that is going to have a scenario applied to it and how the trajectory of data will appear when interacting with a scenario. For example, a determination will need to be made on what asset classes are incorporated and whether those asset classes are identified collectively or separately. There will also have to be a decision on whether to include only publicly traded securities or also private placements. It is more difficult to automate the inclusion of private placements because they do not have security labels and therefore may have to be identified manually. Moreover, there will have to be decisions about what sectors to include and how to identify assets within those sectors. Some sectors may have limited coverage or not have a lot of alternative investment options. Therefore, findings for certain sectors may be more reliable than findings for other sectors. Also, depending on what third-party data products are used, certain assets will be classified as a type of security in some instances but not in other instances. Further, users may only wish to upload a portion of their investment portfolios, such as a region of investments corresponding with a specific scenario or the demands of an agency monitoring or seeking investment reporting.

There are also important differences between transition and physical risk scenario analysis. To begin with, they tend to use different types of scenarios. Transition risk scenarios tend to show plausible assumptions about the deployment of sustainable energy technologies and the development of regulatory policies that will limit greenhouse gas emissions in line with these policies (TCFD 2017b, 12–13). With transition risk modeling, if other factors are held to be the same, it is more difficult for an investment portfolio to be aligned with a lower temperature target. It is also possible that with a very low temperature target, certain industries may not be able to operate and new industries such as those connected with sinks may need to develop, further compounding issues with portfolio alignment.

Physical risk scenarios tend to show the results of projected levels of greenhouse gas emissions on the world's climate that are downscaled to potential local-level impacts (TCFD 2017b, 12–13). With physical risk modeling, if other factors are held to be the same, facilities and property holdings as well as investment portfolios are likely to experience the greatest degrees of risk with higher temperature projections. Moreover, it is possible that with certain high temperatures, there could be cascading effects related to major weather–induced catastrophes, such as the melting of the polar ice caps or bleaching of major coral reefs.

The interaction between transition risk modeling and physical risk modeling is shown in Figure 6.5. When there are lower temperature targets, there are greater transition risks and lower physical risks. Conversely, when there are higher temperature targets, there are lower transition risks and greater physical risks. Therefore, it is important for modelers to use different scenarios when evaluating their assets. As explained by the TCFD, "Given the number of variables and analytical approaches to scenario analysis, there can be a wide range of scenarios that describe various outcomes" (TCFD 2017b, 8). In order to perform the analysis, multiple assumptions need to be made.

A number of companies are familiarizing themselves with transition and physical risk scenario analysis tools. The use of transition risk scenario analysis is starting to become widely used. For instance, 2 Degrees Investing Initiative says that hundreds of companies have uploaded their investment portfolios into their and PRI's free-to-use version of the PACTA tool. The use of physical risk scenario analysis is expanding, too. However, how companies are using transition and physical risk scenario analysis is not entirely clear.

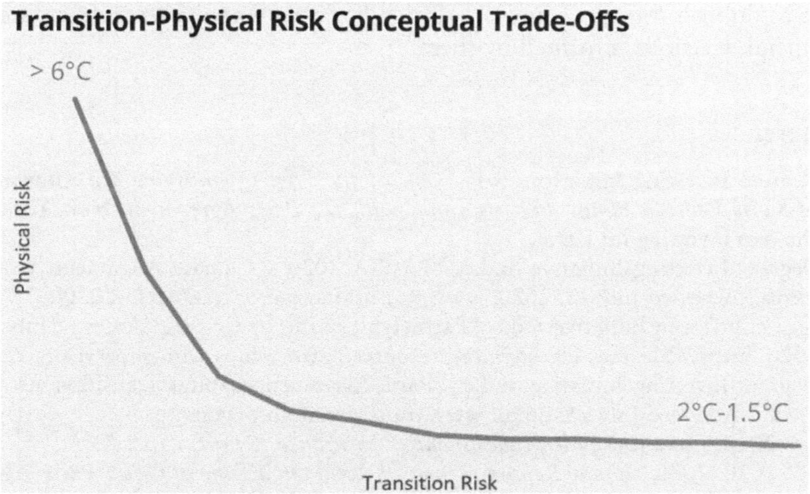

Figure 6.5 Transition-Physical Risk Conceptual Trade-offs

Source: Task Force on Climate-related Financial Disclosures

Outputs from transition and physical scenario analysis can be used to help companies report in jurisdictions where climate change reporting is required, such as in France. Specifically, the French Energy Transition Law requires companies to make climate disclosures but does not currently strictly prescribe how this is to take place, making publication of scenario analysis results on a company website presumably an acceptable option. The PACTA tool can similarly help companies comply with the TCFD recommendations (PRI n.d.a). Four Twenty Seven also explains that its physical risk data can be used for French Energy Transition Law and TCFD reporting.

Outputs from transition and physical risk scenario analysis can also be used to inform investment decisions. Some companies have already said that they will use transition risk scenario analysis testing to help them modify their investment strategy to align with the Paris Agreement (ING n.d.). Other companies are doing the same or stating that they are exploring the possibility of doing so (AXA 2018, 8–11, 2020, 16). Meanwhile, it is possible that at least certain asset managers could be required to at least report findings from such analysis in the relatively near future, especially if they are a pension fund responsible for paying dividends in the long term or any type of asset manager operating in France, where climate-related financial legislation is developing. Along these lines, this scenario analysis can also help inform the decision making of pension fund managers who are required to align with the Paris Agreement objectives or companies that may be required to make similar calculations. However, absent a requirement to either report on carbon risks and green finance or align a company's investment portfolio with a prescribed set of standards, there may not be significant motivation to publish transition and physical risk scenario analysis results or make financial decisions informed by them.

References

2 Degrees Investing Initiative. 2017. *Out of the Fog: Quantifying the Alignment of Swiss Pension Funds and Insurance with the Paris Agreement*. New York: 2 Degrees Investing Initiative.

2 Degrees Investing Initiative. n.d.a. "PACTA 2020 – Climate Alignment Assessment." Accessed July 25, 2020. www.transitionmonitor.com/pacta-2020/.

2 Degrees Investing Initiative. n.d.b. "Partnerships with Supervisors." Accessed July 25, 2020. https://2degrees-investing.org/resource/partnerships-with-supervisors-2/.

2 Degrees Investing Initiative. n.d.c. "Paris Agreement Capital Transition Assessment." Accessed July 25, 2020. www.transitionmonitor.com/.

AXA. 2018. *Climate-Related Investment & Insurance Report*. Paris: AXA.

AXA. 2020. *2020 Climate Report: Renewed Action in a Time of Crisis*. Paris: AXA.

Bank of England. 2019. *General Insurance Stress Test 2019: Scenario Specification, Guidelines and Instructions*. London: Bank of England.

Bank of France. 2020. *Climate-Related Scenarios for Financial Stability Assessment: An Application to France*. Working Paper Series no. 774. Paris: Bank of France.

Bolton, Patrick, Morgan Despres, Luiz Awazu Pereira da Silva, Frédéric Samama, and Romain Svartzman. 2020. *The Green Swan: Central Banking and Financial Stability in the Age of Climate Change*. Basel: Bank of International Settlements and Bank of France.

California Department of Insurance. n.d. "Scenario Analysis." Accessed July 25, 2020. https://interactive.web.insurance.ca.gov/apex_extprd/f?p=250:70.

California Department of Insurance and 2 Degrees Investing Initiative. 2018. *2 Degrees Scenario Analysis, Insurance Companies Operating in California*. Sacramento: California Department of Insurance and 2 Degrees Investing Initiative.

California Department of Insurance and 2 Degrees Investing Initiative. 2019. *2 Degrees Scenario Analysis, Insurance Companies Operating in California*. Sacramento: California Department of Insurance and 2 Degrees Investing Initiative.

Carbon Tracker. 2019. *Measuring Commitment to Climate Goals*. London: Carbon Tracker.

Climate Action 100+. 2019. "2019 Progress Report."

Dent, Kieran, Ben Westwood, and Miguel Segoviano. 2016. *Stress Testing of Banks: An Introduction*. London: Bank of England.

European Union Open Data Portal. n.d. "Flood Hazard Map of the World – 100-Year Return Period." Accessed July 25, 2020. https://data.europa.eu/euodp/en/data/dataset/jrc-floods-floodmapgl_rp100y-tif.

Flavelle, Christopher. 2019. "Moody's Buys Climate Data Firm, Signaling New Scrutiny of Climate Risks." *The New York Times*, July 24. www.nytimes.com/2019/07/24/climate/moodys-ratings-climate-change-data.html.

Four Twenty Seven. 2018. *Using Climate Data, A Primer to Inform the Use of Climate Data in Financial Institutions, Business and Governments*. Berkeley: Four Twenty Seven.

Four Twenty Seven. 2019. *Demystifying Climate Scenario Analysis for Financial Stakeholders*. Berkeley: Four Twenty Seven.

Greenpeace. n.d. "Energy [R]evolution Report Details How United States Can Transition to Nearly 100% Renewable Energy." Accessed May 12, 2014. www.greenpeace.org/usa/news/energy-revolution-report-details-how-us-can-transition-to-nearly-100-renewable-energy/.

ING. n.d. "Terra Approach." Accessed July 25, 2020. www.ing.com/Sustainability/Sustainable-business/INGs-Terra-approach.htm.

Intergovernmental Panel on Climate Change. n.d. "The Intergovernmental Panel on Climate Change." Accessed July 25, 2020. www.ipcc.ch/.

International Energy Agency. 2018. *World Energy Model Documentation*. Paris: International Energy Agency.

International Renewable Energy Agency. n.d. "REmap – Renewable Energy Roadmaps." Accessed July 25, 2020. www.irena.org/remap.

ISO. n.d. "FireLine Reports." *ISO*. Accessed July 25, 2020. www.iso.com/iso PassportHelp/fireline_reports.htm.

Leaton, James, and Andrew Grant. 2017. *2 Degrees of Separation: Transition Risk for Oil and Gas in Low Carbon World*. London: Carbon Tracker.

Mattison, Richard. 2018. "Trucost and S&P Global." *Trucost ESG Analysis and S&P Global*, October 25. www.trucost.com/trucost-blog/trucost-and-sp-global/.

Mills, Evan, Ted Lamm, Sadaf Sukhia, Ethan Elkind, and Aaron Ezroj. 2018. *Trial by Fire: Managing Climate Risks Facing Insurance in the Golden State*. Sacramento: California Department of Insurance.

MSCI. n.d. "The Global Industry Classification Standard." Accessed July 25, 2020. www.msci.com/gics.

MSCI and Carbon Delta. 2019. "Climate Value-at-Risk." Accessed July 25, 2020. www.carbon-delta.com.

Network for Greening the Financial System. 2020. *Guide to Climate Scenario Analysis for Central Banks and Supervisors*. Paris: Network for Greening the Financial System.

Pierce, David W., Daniel R. Cayan, and Laurel Dehaan. 2016. *Climate Projections to Support the 4th California Climate Assessment*. La Jolla: Scripps Institution of Oceanography.

Piotrowski, Jan. 2018. "Markets May be Underpricing Climate-Related Risk." *The Economist*, May 24. www.economist.com/finance-and-economics/2018/05/24/markets-may-be-underpricing-climate-related-risk.

Principles for Responsible Investment. n.d.a. "Launch of First Online and Free Climate Scenario-Based Analysis Tool." Accessed September 3, 2018. www.unpri.org/news-and-press/launch-of-first-online-and-free-climate-scenario-based-analysis-tool/3571.article.

Principles for Responsible Investment. n.d.b. "Climate Scenario Analysis." Accessed July 25, 2020. www.unpri.org/climate-change/directory-of-climate-scenario-tools/3606.article.

Task Force on Climate-related Financial Disclosures. 2017a. *Recommendations of the Task Force on Climate-related Financial Disclosures*. Basel: Task Force on Climate-related Financial Disclosures.

Task Force on Climate-related Financial Disclosures. 2017b. *The Use of Scenario Analysis in Disclosure of Climate-Related Risks and Opportunities*. Basel: Task Force on Climate-related Financial Disclosures.

Thomä, Jakob, and Hugues Chenet. 2019. "Transition Risks and Market Failure: A Theoretical Discourse on Why Financial Models and Economic Agents May Misprice Risk Related to the Transition to a Low-Carbon Economy." In *Stranded Assets: Developments in Finance and Investment*, edited by Ben Caldecott, 45–63. New York: Routledge.

Trucost ESG Analysis and S&P Global. 2019a. *Carbon Earnings at Risk*. London: Trucost ESG Analysis and S&P Global.

Trucost ESG Analysis and S&P Global. 2019b. *Integrating Future Carbon Price Risk into Portfolio Analysis*. London: Trucost ESG Analysis and S&P Global.

Trucost ESG Analysis and S&P Global. 2019c. *TCFD Scenario Analysis: Integrating Future Carbon Price Risk into Portfolio Analysis*. London: Trucost ESG Analysis and S&P Global.

Vermeulen, Robert, Edo Schets, Melanie Lohuis, Barbara Kölbl, David-Jan Jansen, and Willem Heeringa. 2018. *An Energy Transition Risk Stress Test for the Financial System of the Netherlands*. Amsterdam: Dutch Bank.

Vermeulen, Robert, Edo Schets, Melanie Lohuis, Barbara Kölbl, David-Jan Jansen, and Willem Heeringa. 2019. *The Heat Is on: A Framework Measuring Financial Stress Under Disruptive Energy Transition Scenarios*. Amsterdam: Dutch Bank.

World Resources Institute. n.d. "Identify and Evaluate Water Risks Around the World." Accessed July 25, 2020. www.wri.org/aqueduct.

7 Stress testing

Stress testing for liabilities can help insurers and banks address future carbon risks and better inform policymakers.

Banks and insurers generally make money in two ways, investing the capital that they are holding and retaining payments from their customers. A major economic shift, such as a transition to a low-carbon economy, could have major implications on their investments. Major physical changes in the world's climate system could increase the frequency and severity of weather-related catastrophes. The frequency and severity of these catastrophes has important implications for the financial sector, in particular for the underwriting of insurance premiums and for the lending of capital for residential and commercial property. If not otherwise addressed, the increased frequency and severity of weather-related catastrophes is likely to result in consumers making more claims on policies and property owners defaulting on their loans, thereby impacting the ability to collect and retain payments from customers.

7.1 Weather-related catastrophes

In fact, the impact of physical changes in the world's climate system is already becoming observable. Temperature increases in the world's atmosphere are leading to increases in the frequency and severity of certain weather-related catastrophes. For instance, an IPCC study showed that climate change will result in weather-related catastrophes becoming normal occurrences (IPCC 2001; Mills et al. 2018, 2). The World Economic Forum further identified extreme weather events as their top concern (WEF 2018; Mills et al. 2018, 4). Moreover, according to the TCFD, "The scientific consensus is that a continued rise in average global temperatures will have a significant effect on weather-related natural catastrophes and will account for an increasingly large share of natural catastrophe losses" (TCFD 2017, 28). Figure 7.1 shows how temperature increases can exacerbate the frequency and severity of weather-related catastrophes.

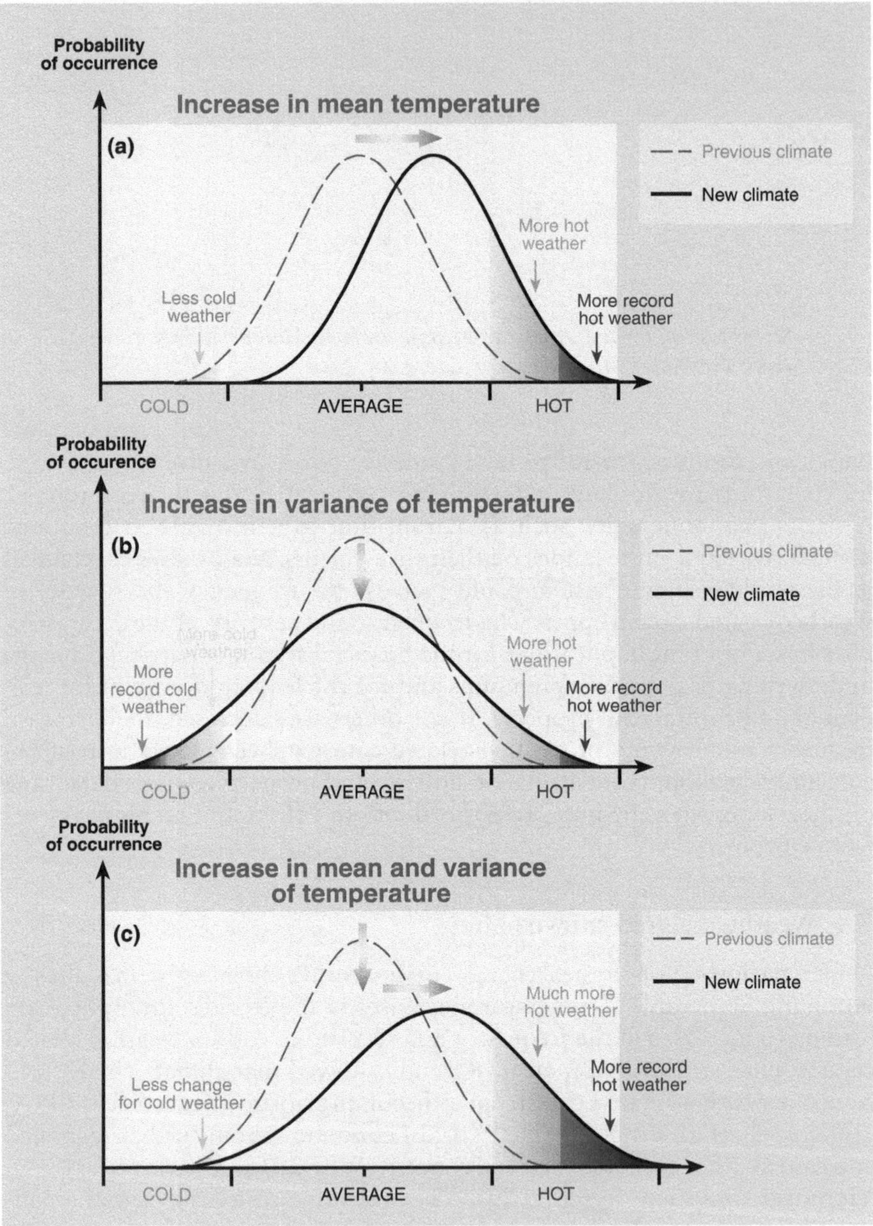

Figure 7.1 Weather-related Catastrophes

Source: Intergovernmental Panel on Climate Change

Wildfires, hurricanes, windstorms, floods, and droughts are among the weather-related catastrophes that are likely to increase in frequency and severity as a result of climate change (IPCC 2001). Significant weather-related losses are already being observed. California has experienced fifteen of its twenty largest wildfires in state history since 2000, with devastating wildfires in the last three consecutive years. The larger and more common wildfires have coincided with warmer, drier, and longer-lasting wildfire seasons. CalFire has said that "climate change is considered a key driver of this trend" (Samenow and Freedman 2019). Total financial losses have amounted to tens of billions of dollars (Samenow and Freedman 2019), and the insurance industry has paid many billions on dollars in claims related to the wildfires in the last few years alone, with over twelve billion dollars in insured losses in a single month (CDI 2019).

Meanwhile, some of the most destructive storms have recently occurred. Harvey, Irma, and Maria hit the Southeastern United States and Caribbean in 2017. In August, Harvey was a Category 4 storm that struck Texas with torrential rainfall and flooding. In September, Irma hit the Virgin Islands and then Florida. Maria hit Puerto Rico, disabling large portions of the island. This set of storms coincided with warmer ocean waters and a conductive Atlantic weather pattern. Harvey, Irma, and Maria are each among the top five costliest storms (Fritz 2018). Total financial losses have accounted to over two hundred billion dollars (Fritz 2018), and the insurance industry has paid between fifty and one hundred billion dollars in claims related to these three storms alone (III n.d.).

7.2 Metrics

Stress testing is an analytical tool used to evaluate the impact of severe events on the profitability or solvency of companies. For stress testing, catastrophe modeling is typically used for evaluating losses connected with weather-related catastrophes. Catastrophe models are computerized systems that estimate potential losses from extreme weather event simulations. Insurers and banks have been using catastrophe models to evaluate their exposure to risks. For insurers, it is important to ensure that they set capital requirements appropriately so that they have sufficient available reserves to pay claims made on policies. In fact, according to the professional service firm PricewaterhouseCoopers, United States insurance regulatory best practices demand solvency assessments to determine that insurance companies have sufficient capital reserves under both normal and stressed conditions (PwC 2016; Mills et al. 2018, 64). For banks, it is also important to ensure that they set capital requirements appropriately so that they make loans in a manner where they are not overleveraged. Catastrophe models can inform these decisions.

7.2.1 Overview

The Bank of England has long advocated for stress testing, specifying best practices and conducting its own stress testing in certain circumstances. This stress testing helps identify events that are especially likely to adversely impact capital requirements and the companies that are likely most exposed to these events (Dent, Westwood, and Segoviano 2016). However, outside of the United Kingdom, insurers and banks have largely been left to decide on their own on when and how to apply stress testing. Financial supervisors and regulators are generally neither specific in how stress testing should take place, nor do they conduct their own stress testing to verify or otherwise assess private-sector practices.

There is very little data on how the private sector is performing stress testing related to extreme weather events. With respect to what types of weather-related stress testing efforts the insurance industry is performing on their own underwriting risks, not much is known on which companies are even conducting such an analysis. In the results of the only mandatory insurance industry–specific climate-related reporting framework, the NAIC Climate Risk Disclosure Survey, not that many insurers state that they are conducing such an analysis. Those insurers that state that they are conducting such an analysis typically do not elaborate on what the analysis entails (Mills et al. 2018, 63–65). Although unfortunately a comprehensive understanding of stress testing practices is not available, insurers and banks do appear to be incorporating catastrophe modeling into business practices in several ways, albeit generally only for single events in specific geographic regions rather than multiple events that could occur in combination (Mills et al. 2018, 14–19, 60–65).

First, insurers and banks retain third-party catastrophe modeling firms to assess their particular portfolios of liabilities for specific events. Doing so is likely less expensive than developing their own tools, but they may not be as precisely calibrated to portfolios of liabilities, and they would not provide a competitive advantage over other companies that are purchasing the same or similar tools. A number of commercial vendors provide catastrophe models, and the modeling of specific simulations can be performed by the vendor or subcontracted to another party.

Second, some insurers and banks are developing their own catastrophe models or running their own simulations on open source catastrophe models. For example, Metropolitan Life Insurance Company tests for storm conditions ranging from 1-in-100 to 1-in-2,000-year probabilities (MetLife 2016; Mills et al. 2018, 64–65). Companies that develop their own catastrophe models or simulations can specifically assess risks that they believe are most pertinent to their portfolios of liabilities and can potentially gain competitive advantages over other companies. However, developing catastrophe models and simulations can be expensive, and the ability of such tools to accurately gauge risk is not certain, especially when compared to catastrophe models and simulations developed by companies specializing in creating catastrophe models and simulations.

Third, some insurers are relying on reinsurers to use catastrophe models when the reinsurers decide whether they will ultimately hold selected layers of insurers' risks for losses related to extreme weather events. Catastrophe models generate pricing for specific loss events using simulations. Insurers and reinsurers then negotiate pricing for specific perils using what catastrophe models reveal about the likelihood of the occurrence of events and related damage amounts. Because reinsurers have worldwide portfolios, this allows for the spreading of risk and less concentrated exposure (American Academy of Actuaries 2018, 26–28).

Fourth, some insurers are providing their portfolio of liabilities to rating agencies that rate companies at least in part on information from catastrophe models (S&P Global 2018; Lavakare and Mawk 2008). A rating using input from catastrophe models allows investors to view a comprehensive evaluation of solvency that takes into account exposure to extreme weather events without insurers or banks having to publicly reveal data that it deems private or proprietary. For example, a rating agency can seek catastrophe model information for financial strength ratings, such as information on exposure to different perils in different geographic regions. However, the majority of insurers and banks are not necessarily providing their underwriting and lending information to rating agencies for evaluation. How such ratings are being composed using such information is not entirely clear either.

There has also not been a lot of sectorwide catastrophe modeling that explores how large catastrophic events could impact sectors as a whole. For a very long time, there had been only limited attempts to test sectorwide impacts of major catastrophes. The most well-known insurance industry stress testing effort was performed over thirty years ago when an insurance research organization financed a report examining how two seven-billion-dollar hurricanes would impact the underwriting of the insurance sector as a whole. The analysis revealed that a number of smaller insurers would likely become insolvent under this simulation being stress tested (All-Industry Research Advisory Council 1986). At that time, such an occurrence of events seemed exceedingly unlikely. However, the likelihood of an occurrence of such events is no longer questioned, but not that much more has been done to test for them on a sectorwide basis (Mills et al. 2018, 63).

Only recently has more sectorwide analysis started to occur. The most significant of this is likely the analysis that the Bank of England has been including in its annual insurer stress tests. In addition to examining the impact of a deterioration in economic markets, the Bank of England has incorporated stress tests for liability shock into its annual insurance sector stress testing. This includes extreme weather event simulations involving major hurricanes, windstorms, and floods. These simulations are to be applied while at the same time taking into consideration a deterioration in market conditions that includes a downward shift in risk-free interest rates, a widening in corporate bond spreads, and a decrease in the value of other assets (Bank of England 2019, 4).

Among the simulations is one where there is a cluster of three hurricanes making landfall in the continental United States with approximately one hundred and eighty-one billion dollars in aggregate insurance sector losses. This stress simulation involves a situation similar to when Harvey, Irma, and Maria occurred in the same year. Here an Irma-like hurricane makes two landfalls in Florida, a Harvey-like hurricane makes landfall in Houston, and a third hurricane dissimilar from Maria makes landfall on the East Coast of the United States. The simulation provides separate modeled hurricane tracks. There is also data provided on how different vendors' catastrophe models display storm severity and gross losses. Insurers are asked to provide loss information specific to types of perils and their lines of business and coverage (Bank of England 2019, 4, 13–16).

Among the extreme weather event simulations is also one where there is a large United Kingdom windstorm and a United Kingdom flood, leading to approximately twenty-five billion dollars of aggregate United Kingdom insurance sector losses. This simulation includes a set of two events. First, there is a large United Kingdom windstorm, resulting in sizable storm surge losses along the East Coast of England. Second, there is significant flooding in England and Wales. There is also data provided on how different vendor catastrophe models display event severity and gross losses. Insurers are asked to provide loss information specific to types of perils and their lines of business and coverage (Bank of England 2019, 23–25).

Some significant sectorwide stress testing efforts have also been conducted in the United States. In 2015, Florida's insurance regulator asked insurers for catastrophe model results that evaluated their ability to absorb one or more historical hurricane simulations. Insurers needed to demonstrate how the simulations would impact their reserves in terms of capital and surplus (FLOIR 2019; Mills et al. 2018, 65). With respect to what types of climate-related catastrophes are being modeled on lending, a study involving Risk Management Solutions – a major international risk modeler – was conducted on the impact of droughts. After applying a drought stress-test model to over a dozen industry sectors in countries around the world, the study determined that there was the possibility of the erosion of creditworthiness with bank loans and increased default risks in certain industries such as water supply, agriculture, food production, and power generation (Carter and Moss 2017, 8–9; Mills et al. 2018, 63–64).

In recent years, there have also been calls to update stress testing to include climate-related projections that can be incorporated into catastrophe models. As far as what is understood from publicly available information, most catastrophe models have incorporated weather events with behaviors based on historic events, but financial supervisors and regulators, academic institutions, companies, and investors are increasingly seeking the inclusion of weather events based on climate projections.

7.2.2 Bank of England

The Bank of England has been leading the development of forward-looking stress tests for liabilities. In 2019, the Bank of England asked insurers to conduct what it labeled as "exploratory exercise" with their annual insurance sector stress testing. The Bank of England explained that the exercise was "designed to capture information about how different firms are exposed to difficult-to-assess risks, eg climate change" (Rule 2019).

The Bank of England outlined several potential climate change scenarios for insurers to evaluate. First, a scenario where there is a sudden transition to a low-carbon economy with global temperature targets being aligned with the Paris Agreement by 2022. Second, a scenario where there are disorderly transitions that are identified in the IPCC Fifth Assessment Report published in 2014. This scenario envisions a long-term transition to a low-carbon economy, with global temperature targets being aligned with the Paris Agreement by 2050. This scenario is based on several decade transition scenarios that are identified in the IPCC's recent report emphasizing the importance of limiting global warming to 1.5 degrees Celsius. Third, a scenario where there is a failed transition to a low-carbon economy, where there is global temperature increase in excess of 4 degrees Celsius by 2100 (Bank of England 2019, 28).

For each potential climate change scenario, the Bank of England outlined likely changes in weather-related occurrences and economic activities. This includes information on expected increases in the frequency and severity of certain weather-related catastrophes, including hurricanes, storms, and floods. This also includes information on expected changes in fossil fuel extraction and power generation using both fossil fuels and non-fossil fuel sources such as renewable energy and nuclear power. Using this information, the Bank of England asked insurers to conduct a stress test for their liabilities in combination with a scenario analysis for their assets (Bank of England 2019, 29–37).

The Bank of England recognized that this exploratory exercise is not an exhaustive assessment and that it may be more beneficial for insurers to develop exercises that more closely meet the challenges that they believe that they are likely to encounter. Indeed, the Bank of England encouraged insurers to provide it with information from separate exercises that they have performed (Bank of England 2019, 29). Nevertheless, the Bank of England still saw the value in insurers conducting the Bank of England's forward-looking stress testing exercise (Rule 2019).

The Bank of England also appears positioned to lead the development of forward-looking climate-related stress tests in the banking sector. The Bank of England recently announced that it will include climate simulations with its bank stress testing starting in 2021. Governor Carney stated that "banks will need to establish how their borrowers are managing current and future climate-related risks and opportunities" (Carney 2019; Horton 2019). According to Governor Carney, banks that align themselves with a net zero world "will be rewarded handsomely" and "[t]hose that fail to adapt will cease to exist" (Carney 2019; Horton 2019). Climate simulations will include one for business as usual and another for the ideal transition to a net zero emissions by 2050 that is consistent with United Kingdom and European Union's legislative objectives (Horton 2019).

Perhaps work already conducted by the University of Cambridge could inform Bank of England forward-looking climate-related stress tests in the banking sector. The University of Cambridge recently released a study showing how climate simulations can be incorporated into catastrophe models used by banks in evaluating their real estate lending portfolios. Specifically, the University of Cambridge identified expected severity increases in European winter wind storm and tropical cyclone risk, uploaded that into an open source catastrophe model, and ran that against a dozen portfolios of assets with a market value of over two trillion dollars. The results revealed that under a 4 degrees Celsius temperature increase, United Kingdom residential mortgage portfolios could see an increase in average annual loss from floods of one hundred and thirty percent, and the number of properties at considerable risk of flood could increase by forty percent (Bronwyn et al. 2019, 7, 44–46).

7.2.3 Chinese efforts

The Industrial and Commercial Bank of China was involved with the Risk Management Solutions effort to model the impact of droughts on the banking industry, which determined that there was the possibility of erosion of creditworthiness of bank loans and increased default risks in certain industries such as water supply, agriculture, food production, and power generation (Carter and Moss 2017; Mills et al. 2018, 63–64). The effort included uploading a Chinese portfolio with approximately 2,500 companies in eleven different industries (Carter and Moss 2017, 30). The effort then ran several drought simulations against this portfolio, with droughts lasting two and five years (Carter and Moss 2017, 16). Credit rating models were used to interpret results (Carter and Moss 2017, 24).

The effort revealed the overall impact of projected droughts on the Chinese economy and impacts of droughts on specific industries and geographic regions. Certain simulations revealed only moderate credit impact in most regions but a material risk for individual companies and potentially a systemic risk to corporate lenders. Smaller companies faced a greater impact than larger companies. This was because smaller firms are more sensitive to revenue reductions and cost increases, and their operating sites are more concentrated in highly industrialized areas near major cities, which face the greatest risks among regions (Carter and Moss 2017, 30).

7.2.4 *California Climate Assessment*

Some significant forward-looking stress tests for liabilities have also been conducted in the insurance sector within the United States. As part of the California's Fourth Climate Change Assessment, RAND Corporation released a report on the impact of changing wildfire risks on California's insurance market. For this report, the California Department of Insurance provided data and otherwise supported the work through the provision of advice. Private insurers provided information on claims. The University of California, Merced provided wildfire simulations (Dixon, Tsang, and Fitts 2018, ii) through a wildfire risk tool that was developed with a Cal-Adapt effort organized by the University of California, Berkeley, and California Energy Commission also as part of California's Fourth Climate Change Assessment (Cal-Adapt n.d.; Westerling 2018, 2020). Pairing historical data on climate, vegetation, fire, and population density with regionally downscaled localized constructed analogs for climate projections, the tool provides California wildfire burn simulations through the end of this century using Scripps Institution of Oceanography downscaled climate model outputs for the IPCC's RCP 4.5 Scenario and RCP 8.5 Scenario (Cal-Adapt n.d.; Westerling 2018, 2020).

The RAND Corporation analysis used this information to evaluate the projected impact of wildfires for structures within dozens of zip codes through the middle and end of this century (Dixon, Tsang, and Fitts 2018, 6–14). This showed that structures in certain zip codes were likely to face a significant increase in risk (Dixon, Tsang, and Fitts 2018, 15–29). The analysis then evaluated the likely increase in cost of insurance premium by zip code using wildfire simulations composed from downscaled climate model outputs for the IPCC's RCP 4.5 Scenario and IPCC's RCP 8.5 Scenario (Dixon, Tsang, and Fitts 2018, 55–64). Of note, in a particularly vulnerable area of California, wildfire simulations composed using downscaled climate model outputs for the IPCC's RCP 8.5 Scenario through the middle of the century showed little or no projected change in insurance premium for about one-third of the zip codes in the area, premium increases ranging from twenty to forty percent for approximately twenty percent of the zip codes in the area, and premium increases somewhere between for the remaining zip codes in the area (Dixon, Tsang, and Fitts 2018, 56).

S&P Global further indicated the importance of forward-looking modeling efforts, such as RAND Corporation's analysis, because existing insurer catastrophe modeling was too limited in its ability to fully evaluate changing wildfire conditions (Wilkinson 2019). This area of modeling is rapidly developing. In particular, California has experienced significant wildfire loss since this RAND Corporation analysis was conducted, and the underlying wildfire simulations have further developed. It is unclear how exactly the additional loss and updated simulations would impact similar analysis, but a forward-looking model could take this into account.

7.2.5 Emerging efforts

The TCFD advised that insurers assess aggregate exposure to weather-related catastrophes of their property business by jurisdiction (TCFD 2017, 31). Moody's concluded that multiple simultaneous and correlated carbon risks could exacerbate current volatility levels and adversely impact credit within the insurance industry, especially for smaller or more geographically concentrated insurers (Moody's 2018; Mills et al. 2018, 2–3). In fact, as a result of fire that destroyed a significant portion of small California mountain community, a small insurer that had written a number of policies in that area became insolvent (Koren 2018). Forward-looking climate-related stress testing has the potential to more precisely address these types of challenges, yielding information that would be valuable in informing companies, supervisors and regulators, and investors on what companies are at risk of insolvency when confronted with weather-related catastrophes in certain areas where there are concentrated risks.

Better projections will become available as forward-looking modeling continues to improve. New climate models are increasing their resolution, and the ability to downscale climate model outputs is advancing. Greater resolution can potentially allow for incorporating individual and community-level remediation efforts, such as firebreaks, vegetation management, and home fortifications. Some modelers already assert that they have such capabilities (Zesty.ai n.d.). Further, using downscaled climate model outputs, new weather simulations allow for the inclusion of additional features. This can include, for instance, wind calculations. Wind can exacerbate or reduce the impact of wildfires depending on the wind's intensity and temperature.

Because of the lack of available data with respect to how proprietary catastrophe models operate, it is difficult for anyone to explain with a high degree of accuracy what the best approach is for catastrophe models to incorporate climate-related risks. This could depend on what type of risk is being evaluated. There may be more comfort with incorporating simulations for weather events that seem most likely to occur in the future and that use models or model outputs with the highest resolution. For instance, while certain perils currently may be modeled with low resolutions, it currently may be possible to model other perils, such as floods, with much better resolutions (Bronwyn et al. 2019, 23). Still, there are remaining decisions to be made on how to best incorporate climate-related risks into property-specific hazard scores and whether to allow overall insurance and lending rates to be influenced by climate-related projections rather than historic loss data. If more data becomes transparent, opinions will likely be influenced by the prevalence of climate-related weather disasters moving forward, the transparency and harmonization of tools, the reliability of predictions, and the holistic incorporation of factors that fairly reflect carbon risks.

References

All-Industry Research Advisory Council. 1986. *Catastrophic Losses: How the Insurance System Would Handle Two $7 Billion Hurricanes*. Malvern: All-Industry Research Advisory Council.

American Academy of Actuaries. 2018. *Uses of Catastrophe Model Output*. Washington, DC: American Academy of Actuaries.

Bank of England. 2019. *General Insurance Stress Test 2019: Scenario Specification, Guidelines and Instructions*. London: Bank of England.

Bronwyn, Claire, Andrew Voysey, Rachel Austin, Emma Jillings, Tom Herbstein, and Amy Nicholass. 2019. *Physical Risk Framework: Understanding the Impacts of Climate Change on Real Estate Lending and Investment Portfolios*. Cambridge: University of Cambridge Institute for Sustainability Leadership.

California Department of Insurance. 2019. "Wildfire Insurance Losses from November 2018 Blazes Top $12 Billion." *Press Release*, May 8.

Cal-Adapt. n.d. "Climate Tools." Accessed July 25, 2020. https://cal-adapt.org/tools/.

Carney, Mark. 2019. "TCFD: Strengthening the Foundations of Sustainable Finance." Speech at TCFD Summit 2019, Tokyo, October.

Carter, Laurence, and Stephen Moss. 2017. *Drought Stress Testing: Making Financial Institutions More Resilient to Environmental Risks*. Geneva: Natural Capital Finance Alliance.

Dent, Kieran, Ben Westwood, and Miguel Segoviano. 2016. *Stress Testing of Banks: An Introduction*. London: Bank of England.

Dixon, Lloyd, Flavia Tsang, and Gary Fitts. 2018. *The Impact of Changing Wildfire Risk on California's Residential Insurance Market*. Santa Monica: RAND.

Florida Office of Insurance Regulation. 2019. "Catastrophe Stress Test Analysis." Florida Office of Insurance Regulation. Accessed July 25, 2020. www.floir.com/Sections/PandC/prepared.aspx.

Fritz, Angela. 2018. "Hurricanes Harvey, Irma, Maria and Nate Were So Destructive, Their Names Have Been Retired." *The Washington Post*, April 12. www.washingtonpost.com/news/capital-weather-gang/wp/2018/04/12/hurricanes-harvey-irma-maria-and-nate-were-so-destructive-their-names-have-been-retired/.

Horton, Emily. 2019. "Mark Carney: Banks Should Stress Test Clients for Climate Risk." *Financial News*, October 8. www.fnlondon.com/articles/mark-carney-banks-should-stress-test-clients-for-climate-risk-20191008.

Insurance Information Institute. n.d. "Facts + Statistics: Hurricanes." Accessed July 25, 2020. www.iii.org/fact-statistic/facts-statistics-hurricanes.

Intergovernmental Panel on Climate Change. 2001. *Climate Change 2001: Synthesis Report*. Geneva: Intergovernmental Panel on Climate Change.

Koren, James Rufus. 2018. "Insurer Merced Went Belly up After Camp Fire. Here's What Policyholders Need to Know." *Los Angeles Times*, December 4. www.latimes.com/business/la-fi-merced-insurance-customers-20181204-story.html.

Lavakare, Ajay, and Kenna Mawk. 2008. "Exposure Data Quality." In *A Guide to Catastrophe Modelling*, edited by Ruth Lythe, 12–14. London: RMS.

Metropolitan Life Insurance Company. 2016. *2016 NAIC Survey Results*. Sacramento: California Department of Insurance.
Mills, Evan, Ted Lamm, Sadaf Sukhia, Ethan Elkind, and Aaron Ezroj. 2018. *Trial by Fire: Managing Climate Risks Facing Insurance in the Golden State*. Sacramento: California Department of Insurance.
Moody's. 2018. *Moody's: Climate Change Heightens Key Risks for P&C Insurance, Reinsurance Sectors*. New York: Moody's.
PricewaterhouseCoopers. 2016. *Stress Testing in the Insurance Industry: A '20/80' Opportunity*. London: PricewaterhouseCoopers.
Rule, David. 2019. *Insurance Stress Test 2019*. London: Bank of England.
Samenow, Jason, and Andrew Freedman. 2019. "What's Driving the Historic California High Wind Events, and Worsening the Wildfires." *The Washington Post*, October, 28. www.washingtonpost.com/weather/2019/10/28/whats-driving-historic-california-high-wind-events-worsening-wildfires/.
S&P Global. 2018. *Are Global Reinsurers Ready for Another Year of Active Natural Catastrophes?* New York: S&P Global.
Task Force on Climate-related Financial Disclosures. 2017. *Implementing the Recommendations of the Task Force on Climate-related Financial Disclosures*. Basel: Task Force on Climate-related Financial Disclosures.
Westerling, Leroy. 2018. *Wildfire Simulations for California's Fourth Climate Change Assessment: Projecting Changes in Extreme Wildfire Events with a Warming Climate*. Merced: UC Merced.
Westerling, Leroy. 2020. *Cal-Adapt: Wildfire Simulations for the Fourth California Climate Change Assessment: Projecting Changes in Extreme Wildfire Events with a Warming Climate*. Berkeley: Cal-Adapt.
Wilkinson, Claire. 2019. "California Wildfire Risks Could Spark Reinsurance Price Surge: S&P." *Business Insurance*, August 16. www.businessinsurance.com/article/20190816/NEWS06/912330181/California-wildfire-risks-could-spark-reinsurance-price-surge-S&P.
World Economic Forum. 2018. *Global Risks Report: 2018*. Geneva: World Economic Forum.
Zesty.ai. n.d. "AI for Insurance." Accessed July 25, 2020. https://zesty.ai/.

8 A comprehensive strategy requires analytical tools

In recent years, the global financial community has become increasingly focused on addressing carbon risks and identifying green finance opportunities. Companies have become more familiar with analytical tools, which they have further refined. Analytical tools have grown in sophistication and relevance while incorporating traditional financial practices. Tools now include reporting frameworks that are harmonizing to the TCFD recommendations, carbon footprint methodologies that can examine supply chains, brown and green taxonomies using third-party data products, and a range of scenario analysis and stress testing assessments. Companies are using this information to evaluate their own business operations, and investors are using the information to decide on what companies to invest in.

Analytical tools are also being used in coordination, which was once largely only a theoretical idea. Several years ago, following a Bank of England, Bank of France, and Dutch Bank meeting in Amsterdam, Bank of France Governor Villeroy de Galhau told the *Financial Times*, "What we will need are forward-looking stress tests assessing the comprehensive interaction between climate change and assets and liabilities" (Hook 2018). California's *Trial by Fire* report postulated as to how a potential stress test examining assets and liabilities could operate:

> A hypothetical example would be one or more consecutive years with record-breaking wildfires in the West, coupled with an occurrence such as multiple hurricanes making landfall in major cities along the Gulf Cost, large losses associated with climate change litigation decisions, the introduction of new carbon taxes, or a major downward stock market correction.
>
> (Mills et al. 2018, 63)

These theoretical ideas now, however, are starting to become a reality.

Because investors are interested in learning about carbon risks and green finance opportunities, major rating agencies are developing or acquiring suites of tools to evaluate carbon risks and green finance opportunities. For instance, S&P Global acquired Trucost, which has developed metrics use for brown and green taxonomy classifications (Mattison 2018). Moody's acquired Four Twenty Seven, which has conducted physical risk modeling on assets (Flavelle 2019). These rating agencies are increasingly incorporating such tools into their ratings while asking for information from companies at least confidentially. For example, S&P Global asks reinsurers and insurers for information on written premiums. S&P Global then incorporates this information into ratings (S&P Global 2016). Moody's evaluates regulated electricity and gas utilities at least in part on their energy mix (Moody's 2017, 33).

For more than a decade, government and private-sector partnerships have been developing to address carbon risk, often with the help of the United Nations and investor networks and nonprofit organizations, such as PRI and Ceres. With help from some of the same entities, a government and private-sector investor effort, Climate Action 100+, recently formed to ensure that the largest corporate greenhouse gas emitters take steps to address carbon risk and is using a suite of analytical tools to do so. Climate Action 100+ established a Technical Advisory Group that consists of some of the world's leading climate modelers, including the Transition Pathway Initiative, Carbon Tracker, and 2 Degrees Investing Initiative and the developer of one of the world's leading climate reporting frameworks, CDP. The Transition Pathway Initiative evaluates whether companies have established long-term quantitative emissions targets and committed to using climate scenarios in their planning. CDP tools look at whether companies have signed as supporters of the TCFD recommendations and completed the CDP climate change questionnaire. Carbon Tracker examines whether companies have investments in oil and gas assets that are misaligned with the IEA's Below 2 Degrees Scenario. 2 Degrees Investing Initiative evaluates whether investments in utility and automobile companies align with the IEA's Below 2 Degrees Scenario (Climate Action 100+ 2019, 19–23, 75–76).

Citing the 2015 United Nations Climate Change Conference in Paris as the impetus for its formation, Climate Action 100+ explains that its investor signatories believe that it is important to engage with the companies that they are invested in to ensure more appropriate disclosures and strategies for aligning with the Paris Agreement. Moreover, doing so is consistent with the investor signatories' fiduciary duties. Several hundred investors, collectively managing over thirty-five trillion dollars in assets, have become Climate Action 100+ signatories (Climate Action 100+ n.d.). The collaboration includes some of the largest institutional investors, such as CalPERS. CalPERS helped develop the concept when it realized that a relatively small number of companies were responsible for the majority of greenhouse gas emissions from its investment portfolio (Rundell 2018). Today, Climate Action 100+ engages over one hundred sixty global companies in more than two dozen markets. These companies account for up to eighty percent of global industrial emissions (Climate Action 100+ 2019, 6). According to an article in *The Economist*, "Investors concerned about climate change have never been better organi[z]ed, thanks to Climate Action 100+ . . . [n]or have they ever had more success" (Economist 2019).

Even though the capabilities of analytical tools have increased significantly, there are still additional potential opportunities for improving them. Reporting frameworks could further harmonize with the TCFD recommendations and each other so it is less burdensome for companies to make disclosures. The more that companies can satisfy their reporting obligations for multiple frameworks using the same underlying data and even the same disclosures, the more likely it is that companies will be able to provide appropriate and useful information. Reporting frameworks themselves can be further updated to include more green finance disclosures, especially where companies do not have another readily available means for making such disclosures. With harmonization to the TCFD recommendations, the emphasis has been on disclosing carbon risks with less robust disclosures for green finance. Not all governments may view green finance efforts as material in the same manner as addressing carbon risks. However, governments and companies are embracing green finance efforts, and analytical tools are quickly developing to identify green finance opportunities. Reporting frameworks can also allow for more scenario analysis disclosures. Scenario analysis is discussed in detail in the TCFD recommendations but has developed rapidly. Gathering and sharing more information on best practices would be informative.

Next, methodologies for calculating carbon footprints could further harmonize, and there could be more extensive applications of this analytical tool. Further harmonization of reporting may be especially helpful with supply chain metrics with Scope 3 emissions reporting, especially if granular greenhouse data is available that allows companies to identify opportunities for greenhouse gas reductions that might not be immediately obvious and for which other tools may not provide as much promise. As harmonization in reporting allows for more comparability among footprints, investors may also benefit more from applying carbon weighted averages to their investment portfolios. In turn, the use of carbon weighted averages could help apply pressure on companies to become more conscientious of their carbon footprints if their footprints reflect on entities that are investing in them. It may also be helpful to further develop means of applying carbon weighted averages in a manner that accounts for differences in sectors, as there is wide variation in greenhouse gas emissions among sectors.

Further, there are potential opportunities to advance brown taxonomies for monitoring assets that could potentially become stranded. There are currently a lot of non-trivial differences in how certain fossil fuel assets are defined. Greater harmonization would make it easier for investors to monitor specific fossil fuel investments and subsequently divest from or engage with companies regarding those investments. More coordinated research efforts could also identify concentrations of particular assets, and discussions with investors about those assets could determine the impact on investors if divestment or engagement efforts are pursued, potentially providing information on efforts that would be cost effective. In addition, brown taxonomies could expand to other covered assets, in particular where there is evidence of high production cost and market volatility, such as tar sands.

Moreover, there are potential opportunities to advance green taxonomies for identifying green investment opportunities. Continued communications between the creators of green taxonomies will likely continue to lead to harmonization that will make it easier for companies to make investments that are recognized as green in multiple taxonomies. More coordinated research efforts could address concerns over what types of green investments are truly additional and reveal how green investments are performing, possibly identifying investments that are universally seen as productive from both an environmental and financial standpoint.

In addition, there are potential opportunities to advance scenario analysis tools. More tools can be provided publicly with training. This is something that the TCFD has been working on, for instance. The TCFD identified concerns over disclosing confidential business information as a potential challenge that limits the ability of companies to disclose scenario analysis results (TCFD 2019, 51). It then offered significant guidance on overcoming this challenge to make substantial disclosures of scenario analysis results. To begin with, the TCFD identified the IEA's and IPCC's scenarios that are in the public domain. Next, the TCFD showed how certain companies are using these and other scenarios to evaluate their business operations. Finally, it showed examples of disclosures being made by the companies (TCFD 2019, 62–74). Current scenario analysis tools can also be expanded to cover more sectors and assets, and there can be more room for users to adjust controlling scenarios. There can also be more stakeholder discussions to ensure harmonization among scenario analysis tools and with other tools that use similar data products and climate assumptions. More coordinated research studies could also identify concentrations of at-risk assets for either divestment or engagement and identify green investment opportunities.

There are also potential opportunities to advance stress testing. Stress testing can include additional factors, such as more carbon risks and mitigation efforts. Ideally, stakeholders in different areas could discuss the risks that present the greatest concern and work on selecting simulations that they feel most accurately present those risks. The results of such stress tests could inform companies on how they will perform when presented with such risks. If stress tests take into account mitigation efforts, the tests could also inform policymakers on mitigation efforts that are likely to best address risks.

Table 8.1 lists potential opportunities for advancing analytical tools.

There are additional potential opportunities for using analytical tools in coordination. Except for some limited static modeling that is not easily comparable, transition and physical risk modeling on assets are not fully connected with each other, and stress testing on liabilities is not connected with either. A major reason that the tools are not fully integrated is that they rely on

Table 8.1 Potential Opportunities for Advancing Tools

Reporting Frameworks	*Carbon Footprints*	*Brown Taxonomy*	*Green Taxonomy*	*Scenario Analysis*	*Stress Testing*
Harmonize Further With TCFD	Harmonize Reporting, Especially With Supply Chain	Harmonize Classifications	Harmonize Classifications	Harmonize Metrics	Harmonize Metrics
Allow for More Green Finance Disclosures	Report More Granular Greenhouse Gas Data	Identify More Potentially Stranded Assets	Explore Additionality of Green Investments	Emphasize 1.5°C Scenario Analysis	Include Additional Factors, Such as More Carbon Risks and Mitigation Efforts
Allow for More Scenario Analysis Disclosures	More Carbon Weighted Average Testing	Include More Covered Assets, Such as Tar Sands	Explore Performance of Green Investments	Identify More Potentially Stranded Assets and Green Finance Opportunities	Identify Greatest Carbon Risks and Most Effective Mitigation Efforts
Allow for Reporting in Annual Statements		Share Information on Divestment and Engagement	Reduce Barriers to Green Investment	Increase Coverage, Including More Industries	

different sets of data that are not always publicly available. To apply a transition and physical risk scenario analysis model to assets, a company would upload an investment portfolio or subset of an investment portfolio into a scenario analysis tool. Although it is not necessarily the case in other parts of the world, in the United States, investment portfolio data is largely public for publicly traded companies. The information can be found in annual reports that can be looked up for specific companies or obtained from a company that aggregates the data. To apply stress testing to a portfolio of liabilities, however, a company would need to upload information on insurance premiums or property loans. This information is not in annual reports or otherwise publicly available. It is guarded under assertions of protecting privacy and the propriety of data. Financial supervisors and regulators do not even necessarily have access to all this information. Therefore, because of the lack of publicly available data, it is currently likely impossible to perform integrated stress tests on a company's assets and liabilities unless information is provided voluntarily. Even if the information is provided voluntarily, it may be difficult to compare the data to corporate peers or otherwise verify the data's accuracy, unless a substantial number of companies are providing such data. However, if these challenges could somehow be addressed, perhaps in the future, there could be single platforms where all analytical tools are integrated.

As analytical tools continue to develop through a number of public and private efforts, the tools will become even more valuable. Michael Bloomberg recently commented on yet another announced effort to advance analytical tools. When sixteen insurers announced that they planned to work with the most advanced, forward-looking scenarios, including indicating that they would work with a below 2 degrees Celsius scenario, Michael Bloomberg said, "The pioneering work of this group will pave the way for greater climate risk transparency and climate action by the global insurance industry, and it's great to see that it's consistent with our [TCFD's] recommendations" (UN 2018).

Major carbon risk and green finance efforts already incorporate analytical tools. The United States does not currently have a comprehensive carbon risk and green finance strategy. However, several states, including California, have been managing a reporting framework that requires certain TCFD disclosures for insurers. California also manages brown taxonomies for pensions funds and insurers and scenario analysis for insurers. The federal government has entertained plans for a carbon risk strategy through the ESG Disclosure Simplification Act and the Climate Risk Disclosure Act and a green finance strategy through a Green New Deal. The ESG Disclosure Simplification Act and Climate Risk Disclosure Act envision TCFD disclosures in annual statements, scenario analysis, and stress testing. The Green New Deal emphasizes directing investments to spur economic development but does not clearly articulate whether that will be done through a green taxonomy. While neither states nor the federal government are requiring companies to calculate carbon footprints, some companies are doing so voluntarily. See Table 8.2.

Table 8.2 United States Carbon Risk and Green Finance Strategy

Reporting Frameworks	*Carbon Footprints*	*Brown Taxonomy*	*Green Taxonomy*	*Scenario Analysis*	*Stress Testing*
Several States Managing a Reporting Framework for Insurers	None	California Manages Brown Taxonomies	None	California Managing Scenario Analysis Efforts	None

Europe has been developing a comprehensive carbon risk and green finance strategy, with some efforts being managed both by the European Commission and countries. The European Union requires large companies to make general environmental disclosures in annual reports, and the European Commission provides guidance on doing so. At the same time, France manages a reporting framework that allows for TCFD disclosures, and the United Kingdom is also planning to develop such a framework. The European Commission is managing a green taxonomy and taking efforts to reduce barriers to making green investments. Several countries are managing scenario analysis efforts for assets. The United Kingdom is managing stress tests. While neither countries nor the European Commission are requiring companies to calculate carbon footprints, some companies are doing so voluntarily. Similarly, although not required, some companies are using existing brown taxonomies. The European Commission is also being pressured to manage its own brown taxonomy. See Table 8.3.

Table 8.3 European Carbon Risk and Green Finance Strategy

Reporting Framework	*Carbon Footprints*	*Brown Taxonomy*	*Green Taxonomy*	*Scenario Analysis*	*Stress Testing*
European Union Requiring Environmental Disclosures in Annual Reports and European Commission Providing Guidance	None	None	European Commission Managing a Green Taxonomy	Countries Managing Scenario Analysis	United Kingdom Managing Stress Tests
France Managing a Reporting Framework					

China has also been developing a comprehensive carbon risk and green finance strategy. The Hong Kong stock exchange has been managing a reporting framework, while the country is planning a national reporting framework. China is managing a green taxonomy and requiring companies to make green investments. China is managing scenario analysis and stress testing efforts. While neither subnational entities nor the federal government are requiring companies to calculate carbon footprints, some companies are doing so voluntarily. See Table 8.4.

Table 8.4 Chinese Carbon Risk and Green Finance Strategy

Reporting Framework	*Carbon Footprints*	*Brown Taxonomy*	*Green Taxonomy*	*Scenario Analysis*	*Stress Testing*
Hong Kong Stock Exchange Managing a Reporting Framework	None	None	China Managing Green Taxonomy	China Managing Scenario Analysis	China Managing Stress Testing

Moving forward, comprehensive strategies for addressing carbon risk and identifying green finance opportunities would further benefit from mustering all available analytical tools as these tools continue to develop. At the forefront of addressing carbon risk is the need for mandatory disclosures. Disclosures will help policymakers evaluate how companies are using analytical tools and provide information on the best practices that are taking place. Where possible, disclosures should be public and readily accessible, in a format such as an annual report. Because tools develop rapidly and are used differently over time, reporting frameworks should also allow for flexibility so that information can be fully captured. Adherence to the TCFD recommendations is also prudent, as the most prominent reporting frameworks have harmonized toward these recommendations.

The TCFD found that the number of companies making disclosures is continuing to increase (Bloomberg 2019, i), with many stating that they plan to fully or at least partially implement the TCFD recommendations (TCFD 2019, 51). However, efforts to disclose do not necessarily go far enough. For instance, Climate Action 100+ found that only thirty percent of analyzed companies have formally supported the TCFD recommendations, and more scenario analysis reporting is needed (Climate Action 100+ 2019, 72). The NGFS, which has been pushing for the increased use of analytical tools through its workstreams, called for "all companies issuing public debt or equity as well as financial sector institutions to disclose in line with the TCFD recommendations" (NGFS 2019, 5).

Comprehensive strategies for addressing carbon risk and identifying green investment opportunities should also require companies to report their carbon footprints. Reporting frameworks allow companies to report this information, but no government is requiring it. These strategies should also encourage investors to report the carbon-weighted average of their investment portfolios. While the TCFD recommendations provided information on this metric, there is no indication that it is being widely used.

To ensure transparency, comprehensive strategies for addressing carbon risk and identifying green investment opportunities should involve managing both brown and green taxonomies. Following stakeholder engagement to fully assess best practices and anticipated implications, either brown and green taxonomies should be developed or an existing taxonomy should be endorsed. Absent an important reason to do otherwise, it would be prudent for any new taxonomy to at least model existing efforts, especially with green taxonomies, which are already widely accepted and harmonizing with one another. Doing so will add credibility to any new taxonomy and facilitate harmonization between taxonomies both in management and implementation. To the extent that any new taxonomy would deviate, there should be an explanation for the deviation and comparable stakeholder involvement to existing efforts. Taxonomies should also be used to provide more public information on the assets that investors are holding and actions that are being taken with respect to those assets. Resources should also be dedicated to supporting certainty surrounding green investments. This could include coordinated research efforts evaluating additionality and performance in achieving environmental and financial gains. This could also include the development of financial mechanisms that lower barriers for investments and otherwise help ensure certainty with achieving green objectives.

Comprehensive strategies for addressing carbon risk and identifying green investment opportunities should involve scenario analysis and stress testing. Where otherwise unavailable, companies should be provided with and trained to use scenario analysis tools, especially 1.5 degrees Celsius scenario analysis. There should also be stress testing for banks and insurers using simulations that are developed through stakeholder engagement. While increasing in use, there are still more opportunities to advance these tools, especially in coordination with one another.

In summary, carbon risk and green finance efforts should further work to incorporate analytical tools. A comprehensive carbon risk and green finance strategy should use all available analytical tools. First, the strategy should mandate TCFD disclosures. Second, the strategy should require companies to prepare carbon footprints with supply chain calculations, allowing companies to analyze opportunities to reduce emissions. Third, the strategy should manage or recognize an existing brown taxonomy, providing clarity concerning which assets are believed to be at risk of potentially becoming stranded. Fourth, the strategy should manage or recognize an existing green taxonomy, providing clarity concerning which assets are believed to present desirable green investment opportunities. Fifth, the strategy should include 1.5 degrees Celsius scenario analysis, either managing analysis for companies or helping companies manage their own analysis. Finally, the strategy should require stress testing, either managing analysis for companies or helping companies manage their own stress testing. Table 8.5 lists how analytical tools should be incorporated into a comprehensive carbon risk and green finance strategy.

Table 8.5 Comprehensive Carbon Risk and Green Finance Strategy

Reporting Framework	*Carbon Footprints*	*Brown Taxonomy*	*Green Taxonomy*	*Scenario Analysis*	*Stress Testing*
Mandate TCFD Disclosures	Require Carbon Footprints with Supply Chain	Manage or Recognize a Brown Taxonomy	Manage or Recognize a Green Taxonomy	Require 1.5°C Scenario Analysis	Require Stress Testing

Over the last several years, the global financial community has organized to evaluate carbon risks and green investment opportunities. These efforts were motivated by Governor Carney's speech and Paris Agreement commitments. They were articulated in the broadly supported TCFD recommendations. Now they are quickly being implemented around the world with renewed urgency following the IPCC and United States National Climate Assessment's bleak assessment of climate change impacts. Moreover, even with a global COVID-19 pandemic, governments are embracing the idea of major investments in renewable energy, labeled green deals. To ensure a smooth transition to a low-carbon economy, major strategies for addressing carbon risk and identifying green finance opportunities should muster all available analytical tools as these tools continue to develop.

References

Bloomberg, Michael. 2019. *Foreword to Task Force on Climate-related Financial Disclosures: Status Report*. Edited by Task Force on Climate-related Financial Disclosures, i. Basel: Task Force on Climate-related Financial Disclosures.

Climate Action 100+. 2019. "2019 Progress Report."

Climate Action 100+. n.d. "Global Investors Driving Business Transition." Accessed July 25, 2020. www.climateaction100.org.

The Economist. 2019. "Oil Majors Face Shareholder Resolutions on Climate Change." *The Economist*, May 20. www.economist.com/business/2019/05/30/oil-majors-face-shareholder-resolutions-on-climate-change.

Flavelle, Christopher. 2019. "Moody's Buys Climate Data Firm, Signaling New Scrutiny of Climate Risks." *The New York Times*, July 24. www.nytimes.com/2019/07/24/climate/moodys-ratings-climate-change-data.html.

Hook, Leslie. 2018. "Central Bank Chiefs Sound Warning on Climate Change." *Financial Times*, April, 8. www.ft.com/content/888616d6-3b07-11e8-b7e0-52972418fec4.

Mattison, Richard. 2018. "Trucost and S&P Global." *Trucost ESG Analysis and S&P Global*, October 25. www.trucost.com/trucost-blog/trucost-and-sp-global/.

Mills, Evan, Ted Lamm, Sadaf Sukhia, Ethan Elkind, and Aaron Ezroj. 2018. *Trial by Fire: Managing Climate Risks Facing Insurance in the Golden State*. Sacramento: California Department of Insurance.

Moody's Investors Service. 2017. *Regulated Electric and Gas Utilities*. New York: Moody's.

Network for Greening the Financial System. 2019. *A Call for Action, Climate Change as a Source of Financial Risk*. Paris: Network for Greening the Financial System.

Rundell, Sarah. 2018. "PGGM, CalPERS Detail Climate Policies." *top1000funds.com*, October 3. www.top1000funds.com/2018/10/pggm-calpers-detail-climate-policies/.

S&P Global, RatingsDirect. 2016. *How We Capture Catastrophe Modeling Uncertainty In (Re)insurance Ratings*. New York: S&P Global, RatingsDirect.

Task Force on Climate-related Financial Disclosures. 2019. *Task Force on Climate-related Financial Disclosures: Status Report*. Basel: Task Force on Climate-related Financial Disclosures.

United Nations. n.d. "UNEP FI Working with 16 Global Insurers to Better Understand Risk & Implement TCFD Recommendations." Accessed November 13, 2018. www.unepfi.org/news/industries/insurance/unep-fi-working-with-16-global-insurers-to-better-understand-risk-implement-tcfd-recommendations/.

Appendix A – Glossary of terms

Bank of England: The United Kingdom financial supervisor that has authority over banks and insurers.

Bank of France: The French financial supervisor that has authority over banks and insurers.

Brown Taxonomy: A classification system for identifying fossil fuel assets.

California Department of Insurance: California's insurance regulator.

California Public Employees' Retirement System (CalPERS): California's public employee pension fund that covers most state employees.

California State Teachers' Retirement System (CalSTRS): California's public pension fund for teachers.

Carbon Bubble Hypothesis: A hypothesis explaining that the desire to restrict climate change creates a carbon budget of the amount of emissions that the world can afford which is less than the emissions that would be associated with using the world's fossil fuel reserves.

Carbon Disclosure Project (now CDP): An organization that has developed a reporting framework that collects climate change information and manages the world's largest carbon footprint database.

Carbon Footprint Calculation: An analytical tool used to quantify how much greenhouse gas companies emit into the atmosphere as a direct or indirect result of their operations.

Carbon Risk: The risks presented by an adapting climate and responses to that adapting climate, possibly limited to transition risks in some usages.

Catastrophe Model: A computerized system that estimates potential losses from extreme weather event simulations.

Ceres: A nonprofit organization advocating for a socially responsible economy.

China Banking and Insurance Regulatory Commission: The Chinese financial regulator that has authority over banks and insurers.

Climate Bonds Initiative (CBI): An organization that has a certification program that identifies debt instruments that can appropriately be labeled as green bonds under the Green Bond Principles.

Climate Action 100+: An investor initiative aimed at encouraging companies to reduce their greenhouse gas emissions.

Climate Disclosures Standards Board (CDSB): An organization that has developed a reporting framework that can be used with the SASB standards to make investor-focused financially material disclosures.

Climate Risk Carbon Initiative: A publicly available classification system that identifies fossil fuel investments held by insurers.

Dutch Bank: The Netherlands financial supervisor that has authority over banks and insurers.

Environmental, Social, and Corporate Governance (ESG): A set of factors that are viewed as central in addressing sustainable and ethical corporate practices.

European Commission Directorate-General for Financial Stability (DG FISMA): The government entity overseeing the implementation of financial legislation that impacts the European Union.

European Green Deal: An expansive plan to spur sustainable growth in the European Union using mechanisms established in the European Union Action Plan on Sustainable Finance.

European Union Action Plan on Sustainable Finance: A comprehensive strategy to address sustainability risks and opportunities within the European Union.

European Union Green Bond Standard: A European Union classification system for certifying debt instruments as green bonds.

European Union Taxonomy: An overarching European Union classification system for identifying environmentally sustainable financial activities.

French Energy Transition Law: A law strengthening carbon disclosure requirements for companies operating in France and introducing them for institutional investors.

Fourth Climate Change Assessment: A major climate research collaboration supported by California government agencies and researchers.

GHG Protocol Corporate Standard: Guidance on how companies should report their greenhouse gas emissions.

Global Coal Exit List: A publicly available classification system used to identify corporate securities as thermal coal investments.

Global Financial Community: A number of public and private actors working on various financial matters that intersect with carbon risk and green finance.

Global Reporting Initiative (GRI): An organization that has developed guidelines that companies use to publish sustainability reports.

Green Bond Endorsed Project Catalogue: A Chinese classification system for identifying debt instruments as green bonds.

Green Bond Principles: An international classification system for certifying debt instruments as green bonds.

Green Finance: The effort to mobilize capital for environmentally sustainable activities.

Green New Deal: An expansive plan to spur sustainable growth and create green jobs in the United States.

Green Taxonomy: A classification system for identifying environmentally sustainable activities.

Greenwashing: Inappropriately labeling securities as sustainable or green.

Intergovernmental Panel on Climate Change (IPCC): A United Nations research organization that develops climate scenarios for physical risk analysis.

International Energy Agency (IEA): An international organization that develops climate scenarios for transition risk analysis.

Kyoto Protocol: An international agreement where countries restrict six greenhouse gases.

National Association of Insurance Commissioners (NAIC): A regulatory support organization that is governed by state insurance regulators in the United States.

Network for Greening the Financial System (NGFS): A global organization of bank supervisors and regulators.

One Planet Summit: A summit convened by French President Macron to discuss the best practices for addressing climate change.

Organization for Economic Co-operation and Development (OECD): An organization consisting largely of economically well-developed countries.

Paris Agreement: An international agreement where countries commit to limiting average global temperatures to no more than a 2 degrees Celsius increase over preindustrial levels and further seek a well below 2 degrees Celsius goal with a 1.5 degrees Celsius target.

Paris Agreement Capital Transition Assessment (PACTA): A scenario analysis energy transition risk tool developed by 2 Degrees Investing Initiative with funding from the European Union.

People's Bank of China: The Chinese financial supervisor that has authority over banks.

Physical Risks: The risks associated with increasing greenhouse gas emissions, including the increased frequency and severity of certain weather and non-weather events.

Principles for Responsible Investment (PRI): An organization of investors committed to sustainable investment principles.

Reporting Framework: A structure for disclosing various organizational information such as efforts to address climate risks and opportunities.

Scenario Analysis: An analytical tool that allows for projecting the impacts of climate change based on the modeling of certain possible scenarios.

Stranded Assets: Fossil fuel assets that would become unburnable if the world transitions to a low-carbon economy.

Stress Testing: An analytical tool used to evaluate the impact of severe events on the profitability and solvency of companies.

Sustainability Accounting Standards Board (SASB): An organization that has developed environmental disclosure standards for investor-focused financially material reporting.

Sustainability Development Goals: A set of seventeen sustainability goals developed by the United Nations.

Sustainable Insurance Forum: A global organization of insurance supervisors and regulators.

Task Force on Climate-related Financial Disclosures (TCFD): An international collaboration of government and business leaders that has developed a set of recommendations for reporting and otherwise evaluating climate risks and opportunities.

Transition Risks: The risks associated with a shift to a low-carbon economy.

United Nations Climate Change Conference: An annual gathering of nations to discuss climate commitments.

Appendix B – Timeline

1997	Kyoto Protocol identifies six greenhouse gases to restrict globally
1998	Carbon footprint metrics are published
2002	Initial climate risk reporting framework is developed
2011	Carbon bubble hypothesis is proposed
2013	China pushes for green lending and investments
2014	Primary green bond taxonomy is created
2015	Governor Mark Carney details climate-related financial risks to European business leaders
2015	World leaders commit to restrict temperature increases to no more than 2 degrees Celsius at the United Nations Climate Change Conference in Paris
2015–2016	California pension funds and insurance regulator urge divestment from thermal coal
2016	The French government implements the French Energy Transition Law
2016	Financial supervisors and regulators convene to form a working group to address climate risks in the insurance sector
2017	Task Force on Climate-related Financial Disclosures publishes its recommendations
2017	One Planet Summit highlights tackling financial risks as a critical strategy for addressing climate change and Climate Action 100+ is launched
2017	Financial supervisors and regulators convene to form a working group to address climate risks in the banking sector
2018	Leading climate risk reporting frameworks incorporate Task Force on Climate-related Financial Disclosures recommendations
2018	Scenario analysis is used by banks, insurers, and pension funds to evaluate climate risks with a 2 degrees Celsius target in mind
2018	United Nations Intergovernmental Panel on Climate Change and United States National Climate Assessment reports stress importance of a 1.5 degrees Celsius target
2019	Reporting framework and scenario analysis tools recalibrate to evaluate risks with a 1.5 degrees Celsius target in mind
2019	Bank of England releases climate-related simulations for catastrophe modeling
2019	European Union starts implementing its action plan on sustainable finance
2019	United States considers moving forward with the ESG Disclosure Act, the Climate Risk Disclosure Act, and a Green New Deal
2020	Europe moves forward with a European Green Deal
2020	China considers a national climate disclosure law and dedicates over a trillion dollars to a range of low-carbon technologies

Index

Note: Page numbers in *italics* indicate figures and page numbers in **bold** indicate tables on the corresponding page.